A-Z RELATIVITY

A-Z RELATIVITY

Prof. G.K. Bose

CENTRUM PRESS
NEW DELHI-110002 (INDIA)

CENTRUM PRESS
H.O.: 4360/4, Ansari Road, Daryaganj,
New Delhi-110 002 (India)
Ph.: 23278000, 23261597

B.O.: No. 1015, Ist Main Road, BSK IIIrd Stage
IIIrd Phase, IIIrd Block,
Bangalore - 560 085 (India)
Tel.: 080-41723429
Visit us at: www.centrumpress.com

A-Z Relativity

First Edition, 2009

ISBN 978-93-80106-35-9

PRINTED IN INDIA

Printed at Salasar Imaging Systems, Delhi-110035 (India)

Contents

Preface *vii*

1. First Principles 1
2. The Relativity of Light 30
3. Symmetry of Relativity 91
4. Doppler Shift for Sound and Light 111
5. The Sagnac Effect 147
6. Special Relativity 215
7. The Lorentz Transformations 248

Index 295

Preface

Relativity is the *Geometric Theory of Gravitation*. It is the state-of-the art description of Gravity in modern physics. It unifies *Special Relativity and Newton's Law of Universal Gravitation*, and describes gravity as a property of the geometry of space and time, or space-time. In particular, the curvature of space-time is directly related to the four-momentum of whatever matter and radiation are present. The relation is specified by the *Einstein Field Equations, a system of Partial Differential Equations*.

The predictions of general relativity differ significantly from those of classical physics, especially concerning the passage of time, the geometry of space, the motion of bodies in free fall, and the propagation of light. Examples of such differences include gravitational time dilation, the gravitational redshift of light, and the gravitational time delay. General relativity's predictions have been confirmed in all observations and experiments to date. Although general relativity is not the only relativistic theory of gravity, it is the simplest theory that is consistent with experimental data. However, unanswered questions remain, the most fundamental being how general relativity can be reconciled with the laws of quantum physics to produce a complete and self-consistent theory of quantum gravity.

Einstein's theory has important astrophysical applications. It points towards the existence of black holes—regions of space in which space and time are distorted in such a way that nothing, not even light, can escape—as an end-state for massive stars.

Author

Chapter 1

First Principles

FROM EXPERIENCE TO SPACETIME

We ordinarily take for granted the existence through time of objects moving according to fixed laws in three-dimensional space, but this is a highly abstract model of the objective world, far removed from the raw sense impressions that comprise our actual experience.

This model may be consistent with our sense impressions, but it certainly is not uniquely determined by them. For example, Ptolemy and Copernicus constructed two very different conceptual models of the heavens based on essentially the same set of raw sense impressions.

Likewise Weber and Maxwell synthesized two very different conceptual models of electromagnetism to account for a single set of observed phenomena. The fact that our raw sense impressions and experiences are (at least nominally) compatible with widely differing concepts of the world has led some philosophers to suggest that we should dispense with the idea of an "objective world" altogether, and base our physical theories on nothing but direct sense impressions, all else being merely the products of our imaginations.

Berkeley expressed the positivist identification of sense impressions with objective existence by the famous phrase "*esse est percipi*".

However, all attempts to base physical theories on nothing but raw sense impressions, avoiding arbitrary conceptual elements, invariably founder at the very start, because we have no sure means of distinguishing sense impressions from our thoughts and ideas.

In fact, even the decision to *make* such a distinction represents a significant conceptual choice, one that is not strictly necessary on the basis of experience.

The process by which we, as individuals, learn to recognize sense impressions induced by an external world, and to distinguish them from our own internal thoughts and ideas, is highly complicated, and perhaps ultimately inexplicable. As Einstein put it "the eternal mystery of the world is its comprehensibility".

Nevertheless, in order to examine the epistemological foundations of any physical theory, we must give some consideration to how the elements of the theory are actually derived from our raw sense impressions, without automatically interpreting them in conventional terms. On the other hand, if we suppress *every* pre-conceived notion, including ordinary rules of reasoning, we can hardly hope to make any progress. We must choose a level of abstraction deep enough to give a meaningful perspective, but not so deep that it can never be connected to conventional ideas.

As an example of a moderately abstract model of experience, we might represent an idealized observer as a linearly ordered sequence of states, each of which is a function of the preceding states and of a set of raw sense impressions from external sources. This already entails two profound choices.

First, it is a purely passive model, in the sense that it does not invoke volition or free will. All conditional statements in this model must be interpreted only as correlations, because without freedom it is meaningless to talk about the different consequences of alternate hypothetical actions. Second, by stipulating that the states are functions of the *preceding* but not the subsequent states we introduce an inherent directional asymmetry to experience, even though the justification for this is far from clear.

Still another choice must be made as to whether the sequence of states and experiences is continuous or discrete. In either case we can parameterize the sequence by a variable λ, and for the sake of definiteness we might represent each state $S(\lambda)$ and the corresponding sense impressions $E(\lambda)$ by strings of binary bits.

Now, because of the mysterious comprehensibility of the world, it may happen that some functions of S are correlated with some functions of E. Since this is a passive model by assumption, we cannot assert anything more than statistical correlations, because we do not have the freedom to arbitrarily vary S and determine the resulting E, but in principle we could still passively encounter enough variety of states and experiences to infer the most prominent correlations.

These most primitive correlations are presumably "hard-wired" into higher-level categories of senses and concepts (i.e., state variables), rather than being sorted out cognitively. In terms of these higher-level variables we might find that over some range of λ the sense impressions $E(\lambda)$ are strictly correlated with three functions θ, ϕ, α of the state $S(\lambda)$, which change only incrementally from one state to the next. Also, we may find that E is only incrementally different for incremental differences in θ, ϕ, α (independent of the prior values of those functions), and that this is the smallest and simplest set of functions with this property.

Finally, suppose the sense impressions corresponding to a given set of values of the state functions are identical if the values of those functions are increased or decreased by some constant.

This describes roughly how an abstract observer might infer an orientation space along with the associated modes of interaction. In conventional terms, the observer infers the existence of external objects which induce a particular set of sense impressions depending on the observer's orientation.

At some point the observer may begin to perceive deviations from the simple three-variable orientation model, and find it necessary to adopt a more complicated conceptual model in order to accommodate the sequence of sense impressions.

It remains true that the simple orientation model applies over sufficiently small ranges of states, but the sense impressions corresponding to each orientation may vary as a function of three additional state variables, which in conventional terms represent the spatial position of the observer. Like the orientation variables, these translation variables, which we might label x, y, and z, change only incrementally from one state to the next, but unlike the orientation variables there is no apparent periodicity.

Note that the success of this process of induction relies on a stratification of experiences, allowing the orientation effects to be discerned first, more or less independent of the translation effects. Then, once the orientation model has been established, the relatively small deviations from it (over small ranges of the state variable) could be interpreted as the effects of translatory motion. If not for this stratification (either in magnitude or in some other attribute), it might never be possible to infer the distinct sources of variation in our sense impressions. (On a more subtle level, the detailed *metrical* aspects of these translation variables will also be found to differ from those of the orientation variables, but only after quantitative units of measure and coordinates have been established).

Another stage in the development of our hypothetical observer might be prompted by the detection of still more complicated variations in the experiential attributes of successive states. The observer may notice that while most of the orientation space is consistent with a fixed position, some particular features of their sense impressions do not maintain their expected relations to the other features, and no combination of the observer's translation and orientation variables can restore consistency. The inferred external objects of perception can no longer be modeled based on the premise that their relations with respect to each other are unchanging.

Significantly, the observer may notice that some features vary as would be expected if the observer's own positional state had changed in one way, whereas other features vary as would be expected if the observer's positions had changed in a different way.

From this recognition the observer concludes that, just as he himself can translate through the space, so also can individual external objects, and the relations are reciprocal. Thus, to each object we now assign an independent set of translation coordinates for each state of the observer.

In so doing we have made another important conceptual choice, namely, to regard "external objects" as having individual identities that persist from one state to the next. Other interpretations are possible. For example, we could account for the apparent motion of objects by supposing that one external entity simply ceases to exist, and another similar entity in a slightly different position comes into existence. According to this view, there would be no such thing as motion, but simply a sequence of arrangements of objects with some similarities.

This may seem obtuse, but according to quantum mechanics it actually is not possible to unambiguously map the identities of individual elementary particles (such as electrons) from one event to another (because their wave functions overlap). Thus the seemingly innocuous assumption of continuous and persistent identities for material objects through time is actually, on some level, demonstrably false.

However, on the macroscopic level, physical objects do seem to maintain individual identities, or at least it is possible to successfully model our sense impressions based on the assumption of persistent identities (because the overlaps between wave functions are negligible), and this success is the justification for introducing the concept of motion for the objects of experience.

The conceptual model of our hypothetical observer now involves something that we may call *distance*, related to the translational state variables, but it's worth noting that we have no direct perception of distances between ourselves and the assumed external objects, and even less between one external object and another.

We have only our immediate sense impressions, which are understood to be purely local interactions, involving signals of some kind impinging on our senses. We infer from these signals a conceptual model of space and time within which external objects reside and move.

This model actually entails two distinct kinds of extent, which we may call distance and length. An object, consisting of a locus of sense impressions that maintains a degree of coherence over time, has a spatial length, as do the paths that objects may follow in their motions, but the conceptual model of space also allows us to conceive of a distance between two objects, defined as the length of the shortest possible path between them.

The task of quantifying these distances, and of relating the orientation variables with the translation variables, then involves further assumptions. Since this is a passive model, all changes are strictly known only as a function of the single state variable, but we imagine other pseudo-independent variables based on the observed correlations. We have two means of quantifying spatial distances.

One is by observing the near coincidence of one or more stable entities (measuring rods) with the interval to be quantified, and the other is to observe the change in the internal state variable as an object of stable speed moves from one end of the interval to the other. Thus we can quantify a spatial interval in terms of some reference spatial interval, or in terms of the associated temporal interval based on some reference state of motion. We identify these references purely by induction based on experience.

Combining the rotational symmetries and the apparent translational distances that we infer from our primary sense impressions, we conventionally arrive at a conception of the external world that is, in some sense, the *dual* of our subjective experience. In other words, we interpret our subjective experience as a one-dimensional temporally-ordered sequence of events, whereas we conceive of "the objective world now" corresponding to a single perceived event as a three-dimensional expanse of space as illustrated below:

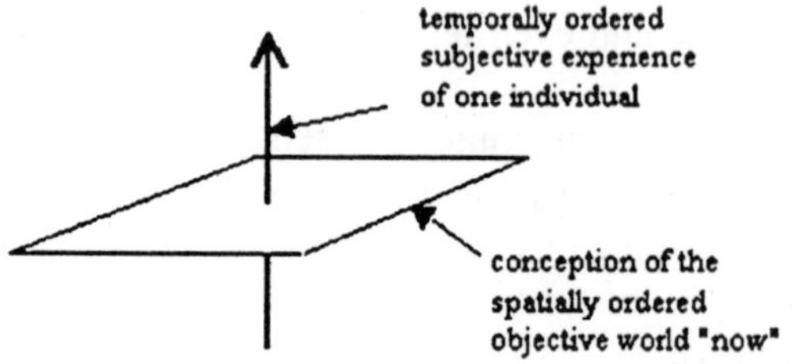

In this way we intuitively conceive of time and space as inherently perpendicular dimensions, but complications arise if we posit that each event along our subjective path resides in, and is an element of, an objective world. If the events along any path are discrete, then we might imagine a simple sequence of discrete "instantaneous worlds":

One difficulty with this arrangement is that it isn't clear how (or whether) these worlds interact with each other. If we regard each "instant" as a complete copy of the spatial universe, separate from every other instant, then there seems to be no definite way to identify an object in one world with "the same" object in another, particularly considering qualitatively identical objects such as electrons.

If we have two electrons assigned the labels A and B in one instant of time, and if we find two electrons in the next instant of time, we have no certain way of deciding which of them was the "A" electron from the previous instant. (In fact, we cannot even map the spatial locations of one instant to "the same" locations in any other instant).

This illustrates how the classical concept of motion is necessarily based on the assumption of persistent identities of objects from one instant to another.

Since it does seem possible (at least in the classical realm) to organize our experiences in terms of individual objects with persistent and unambiguous identities over time, we may be led to suspect that the sequence of existence of an individual or object in any one instant must be, in some sense, connected to or *contiguous* with its existence in neighboring instants.

If these objects are the constituents of "the world", this suggests that *space* itself at any "instant" is *continuous* with the spaces of neighboring instants. This is important because it implies a definite connectivity between neighboring world-spaces, and this, as we'll see, places a crucial constraint on the relativity of motion.

Another complication concerns the relative orderings of world-instants along different paths.

Our schematic above implied that the "instantaneous worlds" are well-ordered in the sense that they are encountered in the same order along every individual's path, but of course this need not be the case.

For example, we could equally well imagine an arrangement in which the "instantaneous worlds" are skewed, so that different individuals encounter them in different orders, as illustrated below:

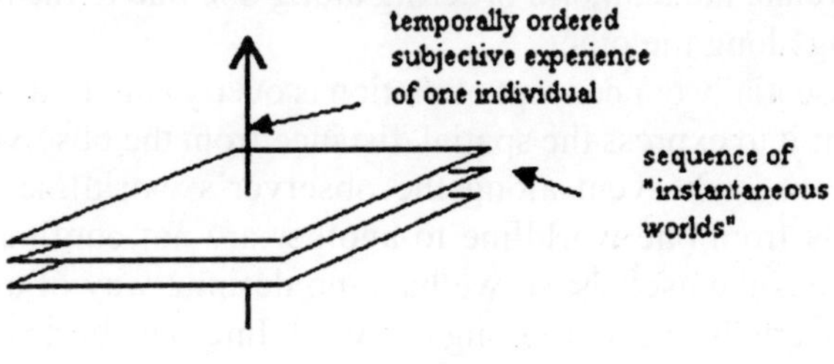

The concept of motion assumes the world can be analyzed in two different ways, first as the union of a set of mutually exclusive "events", and second as a set of "objects" each of which participates in an ordered sequence of events.

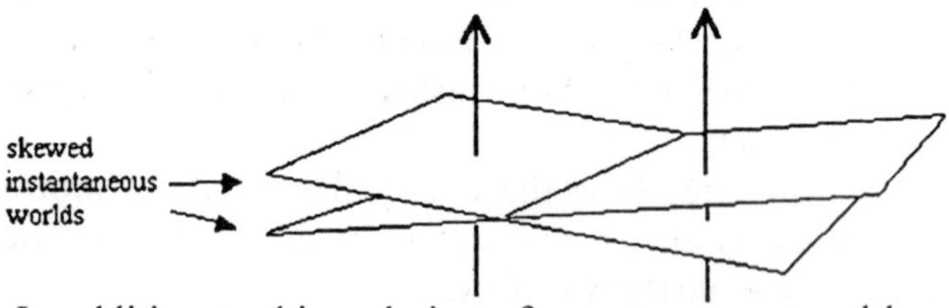

In addition to this ordering of events encountered by each individual object, we must also assume both a co-lateral ordering of the events associated with different objects, *and* a transverse ordering of events from one object to another. These three kinds of orderings are illustrated schematically below.

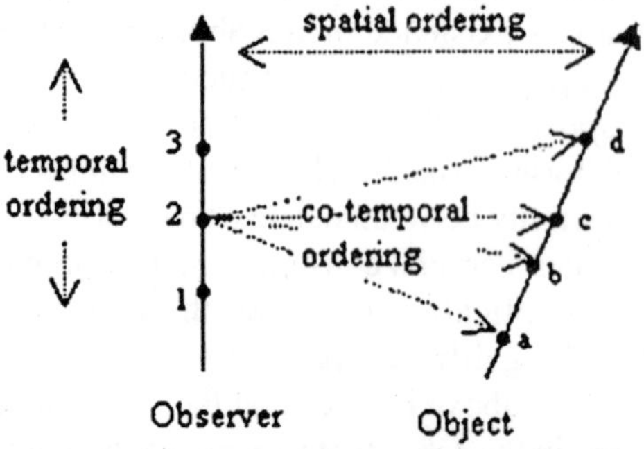

This diagram suggests that the idea of motion is actually quite complex, even in this simple abstract model. Intuitively we regard motion as something like the derivative of the spatial "position" with respect to "time", but we can't even unambiguously define the distance between two worldlines, because it depends on how we correlate the temporal ordering along one line to the temporal ordering along the other.

Essentially our concept of motion is overly ambitious, because we want it to express the spatial distance from the observer to the object for each event along the observer's worldline, but the intervals from one worldline to another are not confined to the worldlines themselves, so we have no definite way of assigning those intervals to events along our worldline. The best we can do is correlate *all* the intervals from a particular point on the observer's worldline to the object's worldline.

When we considered everything in terms of the sense impressions of just a single observer this was not an issue, since only one parameterization was needed to map the experiences of that observer, interpreted solipsistically. Any convenient parameterization was suitable.

When we go on to consider multiple observers and objects we can still allow each observer to map his experiences and internal states using the most convenient terms of reference (which will presumably include his own state-index as the temporal coordinate), but now the question arises as to how all these private coordinate systems are related to each other. To answer this question we need to formalize our parameterizations into abstract systems of coordinates, and then consider how the coordinates of any given event with respect to one system are related to the coordinates of the same event with respect to another system.

Considering how far removed from our raw sense impressions is our conceptual model of the external world, and how many unjustified assumptions and interpolations are involved in its construction, it's easy to see why some philosophers have advocated the rejection of all conceptual models.

However, the fact remains that the imperative to reconcile our experience with *some* model of an objective external world has been one of the most important factors guiding the development of physical theories.

Even in quantum mechanics, arguably the field of physics most resistant to complete realistic reconciliation, we still rely on the "correspondence principle", according to which the observables of the theory must conform to the observables of classical realistic models in the appropriate limits. Naturally our interpretations of experience are always provisional, being necessarily based on incomplete induction, but conceptual models of an objective world have proven (so far) to be indispensable.

SYSTEMS OF REFERENCE

There are many theories of relativity, each of which can be associated with some arbitrariness in our descriptions of events.

For example, suppose we describe the spatial relations between stationary particles on a line by assigning a real-valued coordinate to each particle, such that the distance between any two particles equals the difference between their coordinates.

There is a degree of arbitrariness in this description due to the fact that all the coordinates could be increased by some arbitrary constant without affecting any of the relations between the particles. Symbolically this translational relativity can be expressed by saying that if x is a suitable system of coordinates for describing the relations between the particles, then so is x + k for any constant k.

Likewise if we describe the spatial relations between stationary particles on a plane by assigning an ordered pair of real-valued coordinates to each particle, such that the squared distance between any two particles equals the sum of the squares of the differences between their respective coordinates, then there is a degree of arbitrariness in the description (in addition to the translational relativity of each individual coordinate) due to the fact that we could rotate the coordinates of every particle by an arbitrary constant angle without affecting any of the relations between the particles.

This relativity of orientation is expressed symbolically by saying that if (x, y) is a suitable system of coordinates for describing the positions of particles on a plane, then so is (ax-by, bx + ay) where $a^2 + b^2 = 1$.These relativities are purely formal, in the sense that they are tautological consequences of the premises, regardless of whether they have any physical applicability.

Our first premise was that it's possible to assign a single real-valued coordinate to each particle on a line such that the distance between any two particles equals the difference between their coordinates.

If this premise is satisfied, the invariance of relations under coordinate transformations from x to x + k follows trivially, but if the pairwise distances between three given particles were, say, 5, 3, and 12 units, then no three numbers could be assigned to the particles such that the pairwise differences equal the distances.

This shows that the n(n–1)/2 pairwise distances between n particles cannot be independent of each other if those distances can be encoded unambiguously by just n coordinates in one dimension or, more generally, by kn coordinates in k dimensions. A suitable system of coordinates in one dimension exists only if the distances between particles satisfy a very restrictive condition.

Letting d(A, B) denote the signed distance from A to B, the condition that must be satisfied is that for every three particles A, B, C we have d(A, B) + d(B, C) + d(C, A) = 0. Of course, this is essentially the definition of co-linearity, but we have no a priori reason to expect this definition to have any applicability in the world of physical objects. The fact that it has wide applicability is a non-trivial aspect of our experience, albeit one that we ordinarily take for granted.

Likewise for particles in a region of three dimensional space the premise that we can assign three numbers to each particle such that the squared distance between any two particles equals the sum of the squares of the differences between their respective coordinates is true only under a very restrictive condition, because there are only 3n degrees of freedom in the n(n−1)/2 pairwise distances between n particles.

Just as we found relativity of orientation for the pair of spatial coordinates x and y, we also find the same relativity for each of the pairs x, z and y, z in three dimensional space. Thus we have translational relativity for each of the four coordinates x, y, z, t, and we have rotational relativity for each pair of spatial coordinates (x, y), (x, z), and (y, z).

This leaves the pairs of coordinates (x,t),(y, t) and (z, t). Not surprisingly we find that there is an analogous arbitrariness in these coordinate pairs, which can be expressed (for the x, t pair) by saying that the relations between the instances of particles on a line as a function of time are unaffected if we replace the x and t coordinates with ax − bt and −bx + at respectively, where $a^2 - b^2 = 1$. These transformations (rotations in the x, t plane through an imaginary angle), which characterize the theory of special relativity, are based on the premise that it is possible to assign pairs of values, x and t, to each instance of each particle on the x axis such that the squared spacetime distance equals the difference between the squares of the differences between the respective coordinates.

Each of the above examples represents an invariance of physically measurable relations under certain classes of linear transformations. Extending this idea, Einstein's general theory of relativity shows how the laws of physics, suitably formulated, are invariant under an even larger class of transformations of space and time coordinates, including non-linear transformations, and how these transformations subsume the phenomena of gravity.

In general relativity the metrical properties of space and time are not constant, so the simple premises on which we based the primitive relativities described above turn out not to be satisfied globally. However, it remains true that those simple premises are satisfied locally, i.e., over sufficiently small regions of space and time, so they continue to be of fundamental importance.

The relativities described above are purely formal and tautological, but it turns out that each of them is closely related to a non-trivial physical symmetry. There exists a large class of identifiable objects whose lengths maintain a fixed proportion to each other under the very same set of transformations that characterize the relativities of the coordinates. In other words, just as we can translate the coordinates on the x axis without affecting the length of any object, we also find a large class of objects that can be individually translated along the x axis without affecting their lengths.

The same applies to rotations and boosts. Such changes are physically distinct from purely formal shifts of the entire coordinate system, because when we move individual objects we are actually changing the relations between objects, since we are moving only a subset of all the coordinated objects. (Also, moving an object from one stationary position to another requires acceleration). Thus for each formal arbitrariness in the system of coordinates there exists a physical symmetry, i.e., a large class of entities whose extents remain in constant proportions to each other when subjected individually to the same transformations.

We refer to these relations as physical symmetries rather than physical invariances, because (for example) we have no basis for asserting that the length of a solid object or the duration of a physical process is invariant under changes in position, orientation or state of motion.

We have no way of assessing the truth of such a statement, because our measures of length and duration are all comparative. We can say only that the spatial and temporal extents of all the "stable" physical entities and processes are affected (if at all) in exactly the same proportion by changes in position, orientation, and state of motion. Of course, given this empirical fact, it is often convenient to speak as if the spatial and temporal extents are invariant, but we shouldn't forget that, from an epistemological standpoint, we can assert only symmetry, not invariance.

In his original presentation of special relativity in 1905 Einstein took measuring rods and clocks as primitive elements, even though he realized the weakness of this approach. He later wrote of the special theory. It is striking that the theory introduces two kinds of physical things, i.e.,

- Measuring rods and clocks.
- All other things, e.g., the electromagnetic field, the material point, etc. This, in a certain sense, is inconsistent; strictly speaking, measuring rods and clocks should emerge as solutions of the basic equations (objects consisting of moving atomic configurations), not, as it were, as theoretically self-sufficient entities.

The procedure was justified, however, because it was clear from the very beginning that the postulates of the theory are not strong enough to deduce from them equations for physical events sufficiently complete and sufficiently free from arbitrariness to form the basis of a theory of measuring rods and clocks.

This is quite similar to the view he expressed many years earlier the solid body and the clock do not in the conceptual edifice of physics play the part of irreducible elements, but that of composite structures, which may not play any independent part in theoretical physics. But it is my conviction that in the present stage of development of theoretical physics these ideas must still be employed as independent ideas; for we are still far from possessing such certain knowledge of theoretical principles as to be able to give exact theoretical constructions of solid bodies and clocks.

The first quote is from his Autobiographical Notes in 1949, whereas the second is from his essay on Geometry and Experience published in 1921. It's interesting how little his views had changed during the intervening 28 years, despite the fact that those years saw the advent of quantum mechanics, which many would say provided the very theoretical principles underlying the construction of solid bodies and clocks that Einstein felt had been lacking.

Whether or not the principles of quantum mechanics are adequate to justify our conceptions of reference lengths and time intervals, the characteristic spatial and temporal extents of quantum phenomena are used today as the basis for all such references. Considering the arbitrariness of absolute coordinates, one might think our spatio-temporal descriptions could be better expressed in purely relational terms, such as by specifying only the mutual distances (minimum path lengths) between objects.

Nevertheless, the most common method of description is to assign absolute coordinates (three spatial and one temporal) to each object, with reference to an established system of coordinates, while recognizing that the choice of coordinate systems is to some extent arbitrary. The relations between objects are then inferred from these absolute (thought somewhat arbitrary) coordinates. This may seem to be a round-about process, but there are several reasons for using absolute coordinate systems to encode the relations between objects, rather than explicitly specifying the relations themselves.

One reason is that this approach enables us to take advantage of the efficiency made possible by the finite dimensionality of space. If there were no limit to the dimensionality of space, then we would expect a set of n particles to have n(n–1)/2 independent pairwise spatial relations, so to explicitly specify all the distances between particles would require n–1 numbers for each particle, representing the distances to each of the other particles. For a large number of particles (to say nothing of a potentially infinite number) this would be impractical. Fortunately the spatial relations between the objects of our experience are not mutually independent.

The nth particle essentially adds only three (rather than n–1) degrees of freedom to the relational configuration. In physical terms this restriction can be clearly seen from the fact that the maximum number of mutually equidistant particles in D-dimensional space is D+1. Experience teaches us that in our physical space we can arrange four, but not five or more, particles such that they are all mutually equidistant, so we conclude that our space has three dimensions.

Historically the use of absolute coordinates rather than explicit relations may also have been partly due to the fact that analytic geometry and Cartesian coordinates were invented (by Fermat, Descartes and others) at almost the same time that the new science of mechanics needed them, just as tensor analysis was invented, three hundred years later, at the very moment when it was needed to facilitate the development of general relativity. .

The coordinate systems of Descartes were not merely efficient, they were also consistent with the ancient Aristotelian belief (also held by Descartes) that there is no such thing as empty space or vacuum, and that continuous substance permeates the universe. In this context we cannot even contemplate explicitly specifying each individual distance between substantial points, because space is regarded as a continuum of substance.

For Aristotle and Descartes, every spatial extent is a measure of the length of some substance, not a pure distance between particles as contemplated by atomists. In this sense we can say that the continuous absolute coordinate systems inherited by modern science from Aristotle and Descartes are a remnant of the Cartesian natural philosophy.

Another, perhaps more compelling, reason for the adoption of abstract coordinate systems in the descriptions of physical phenomena was the need to account for acceleration. As Newton explained with the example of a "spinning pail", the mutual relations between a set of material particles in an instant are not adequate to fully characterize a physical situation – at least not if we are considering only a small subset of all the particles in the universe. (Whether the mutual relations would be adequate if all the matter in the universe was taken into account is an open question).

In retrospect, there were other possible alternatives, such as characterizing not just the relations between particles at a specific instant, but over some temporal span of existence, but this would have required the unification of spatial and temporal measures, which did not occur until much later. Originally the motions of objects were represented simply by allowing the spatial coordinates of each persistent object to be continuous single-valued functions of one real variable, the time coordinate.

Incidentally, one consequence of the use of absolute coordinates is that it automatically entails a breaking of the alleged translational symmetry.

We said previously that the coordinate system x could be replaced by x + k for any real number k, implying that every real value of k is in some sense equally suitable. However, from a strictly mathematical point of view there does not exist a uniform distribution over the real numbers, so this form of representation does not exactly entail the perfect symmetry of position in an infinite space, even if the space is completely empty.

The set of all combinations of values for the three spatial coordinates and one time coordinate is assumed to give a complete coordination not only of the spatial positions of each entity at each time, but of *all possible* spatial positions at all possible times. Any definite set of space and time coordinates constitutes a system of reference.

There are infinitely many distinct ways in which such coordinates can be assigned, but they are not entirely arbitrary, because we limit the range of possibilities by requiring contiguous physical entities to be assigned contiguous coordinates. This imposes a definite structure on the system, so it is more than merely a set of labels; it represents the most primitive laws of physics.

One way of specifying an entire model of a world consisting of n (classical) particles would be to explicitly give the 3n functions $x_j(t)$, $y_j(t)$, $z_j(t)$ for j = 1 to n. In this form, the un-occupied points of space would be irrelevant, since only the actual paths of actual physical entities have any meaning. In fact, it could be argued that only the intersections of these particles have physical significance, so the paths followed by the particles in between their mutual intersections could be regarded as merely hypothetical.

Following this approach we might end up with a purely combinatorial specification of discrete interactions, with no need for the notion of a continuous physical space within which entities reside and move. However, the hypothesis that physical objects have continuous positions as functions of time with respect to a specified system of reference has proven to be extremely useful, especially for purposes of describing simple laws by which the observable interactions can be efficiently described and predicted.

An important class of physical laws that make use of the full spatio-temporal framework consists of laws that are expressed in terms of *fields*.

A field is regarded as existing at each point within the system of coordinates, even those points that are not occupied by a material particle. Therefore, each continuous field existing throughout time has, potentially, far more degrees of freedom than does a discrete particle, or even infinitely many discrete particles.

Arguably, we never actually observe fields, were merely observe effects attributed to fields. It's ironic that we can simplify the descriptions of particles by introducing hypothetical entities (fields) with far more degrees of freedom, but the laws governing the behaviour of these fields (e.g., Maxwell's equations for the electromagnetic field) along with symmetries and simple boundary conditions suffice to constrain the fields so that actually do provide a simplification. (Fields also provide a way of maintaining conservation laws for interactions "at a distance"). Whether the usefulness of the concepts of continuous space, time, and fields suggests that they possess some ontological status is debatable, but the concepts are undeniably useful.

These systems of reference are more than simple labeling. The numerical values of the coordinates are intended to connote physical properties of order and measure. In fact, we might even suppose that the sequence of states of all particles are uniformly parameterized by the time coordinate of our system of reference, but therein lies an ambiguity, because it isn't clear how the temporal states of one particle are to be placed in correspondence with the temporal states of another.

Here we must make an important decision about how our model of the world is to be constructed. We might choose to regard the totality of all entities as comprising a single element in a succession of universal temporal states, in which case the temporal correspondence between entities is unambiguous. In such a universe the temporal coordinate induces a total ordering of events, which is to say, if we let the symbol $\leq$ denote temporal precedence or equality, then for every three events a, b, c we have

(i) $a \leq a$
(ii) if $a \leq b$ and $b \leq a$, then $a = b$
(iii) if $a \leq b$ and $b \leq c$, then $a \leq c$
(iv) either $a \leq b$ or $b \leq a$

However, this is not the only possible choice. We might choose instead to regard the temporal state of each individual particle as an independent quantity, bearing in mind that orderings of the elements of a set are not necessarily total. For example, consider the subsets of a flat plane, and the ordering induced by the inclusion relation $\subseteq$. Obviously the first three axioms of a total ordering are satisfied, because for any three subsets a, b, c of the plane we have (i) $a \subseteq a$, (ii) if $a \subseteq b$ and $b \subseteq a$, then $a = b$, and (iii) if $a \subseteq b$ and $b \subseteq c$, then $a \subseteq c$.

However, the fourth axiom is not satisfied, because it's entirely possible to have two sets neither of which is included in the other. An ordering of this type is called a partial ordering, and we should allow for the possibility that the temporal relations between events induce a partial rather than a total ordering. In fact, we have no *a priori* reason to expect that temporal relations induce even a partial ordering. It is safest to assume that each entity possesses its own temporal state, and let our observations teach us how those states are mutually related, if at all. (Similar caution should be applied when modeling the relations between the spatial states of particles).

Given any system of space and time coordinates we can define infinitely many others such that speeds are preserved. This represents an equivalence relation, and we can then define a reference frame as an equivalence class of coordinate systems such that the speed of each object has the same value in terms of each coordinate system in that class. Thus within a reference frame we can speak of the speed of an object, without needing to specify any particular coordinate system. Of course, just as our coordinate systems are generally valid only locally, so too are the reference frames. Purely kinematic relativity contains enough degrees of freedom that we can simply define our systems of reference (i.e., coordinate systems) to satisfy the additivity of velocity. In other words, we can adopt velocity additivity as a principle, and this is essentially what scientists had tacitly done since ancient times. The great insight of Galileo and his successors was that this principle is inadequate to single out the physically meaningful reference systems.

INERTIA AND RELATIVITY

In his treatise "On the Revolution of Heavenly Spheres" Copernicus argued for the conceivability of a moving Earth by noting that every apparent change in place occurs on account of the movement either of the thing seen or of the spectator, or on account of the necessarily unequal movement of both. No movement is perceptible relatively to things moved equally in the same direction - I mean relatively to the thing seen and the spectator.

This is a purely kinematical conception of relativity, like that of Aristarchus, based on the idea that we judge the positions (and changes in position) of objects only in relation to the positions of other objects. Many of Copernicus's contemporaries rejected the idea of a moving Earth, because we do not directly "sense" any such motion.

To answer this objection, Galileo developed the concept of inertia, which he illustrated by a "thought experiment" involving the behaviour of objects inside a ship which is moving at some constant speed in a straight line. He pointed out that among things which all share equally in any motion, [that motion] does not act, and is as if it did not exist... in throwing something to your friend, you need throw it no more strongly in one direction than in another, the distances being equal... jumping with your feet together, you pass equal spaces in every direction...

Thus Galileo's approach was based on a dynamical rather than a merely kinematic analysis, because he refers to forces acting on bodies, asserting that the dynamic behaviour of bodies is homogeneous and isotropic in terms of (suitably defined) measures in any uniform state of motion.

This soon led to the modern principle of inertial relativity, although Galileo himself seems never to have fully grasped the distinction between accelerated and unaccelerated motion. He believed, for example, that circular motion was a natural state that would persist unless acted upon by some external agent. This shows that the resolution of dynamical behaviour into inertial and non-inertial components - which we generally take for granted today - is more subtle than it may appear. As Newton wrote:

> '...the whole burden of philosophy seems to consist in this: from the phenomena of motions to infer the forces of nature, and then from these forces to deduce other phenomena...'

Newton's doctrine implicitly assumes that forces can be inferred from the motions of objects, but establishing the correspondence between forces and motions is not trivial, because the doctrine is, in a sense, circular. We infer "the forces of nature" from observed motions, and then we account for observed motions in terms of those forces.

This assumes we can distinguish between forced and unforced motion, but there is no *a priori* way of making such a distinction. For example, the roughly circular motion of the Moon around the Earth might suggest the existence of a force (universal gravitation) acting between these two bodies, but it could also be taken as an indication that circular motion is a natural form of unforced motion, as Galileo believed. Different definitions of unforced motion lead to different sets of implied "forces of nature".

The task is to choose a definition of unforced motion that leads to the identification of a set of physical forces that gives the most intelligible decomposition of phenomena. By indirect reasoning, the natural philosophers of the seventeenth century eventually arrived at the idea that, in the complete absence of external forces, an object *would* move uniformly in a straight line, and that, therefore, whenever we observe an object whose speed or direction of motion is changing, we can infer that an external force – proportional to the rate of change of motion – is acting upon that object.

This is the principle of inertia, the most successful principle ever proposed for organizing our knowledge of the natural world. Notice that it refers to how a free object "would" move, because no object is completely free from all external forces. Thus the conditions of this fundamental principle, as stated, are never actually met, which highlights the subtlety of Newton's doctrine, and the aptness of his assertion that it comprises "the whole burden of philosophy".

Also, notice that the principle of inertia does not discriminate between different states of uniform motion in straight lines, so it automatically entails a principle of relativity of dynamics, and in fact the two are essentially synonymous.

The first explicit statement of the modern principle of inertial relativity was apparently made by Pierre Gassendi, who is most often remembered today for reviving the ancient Greek doctrine of atomism. In the 1630's Gassendi repeated many of Galileo's experiments with motion, and interpreted them from a more abstract point of view, consciously separating out gravity as an external influence, and recognizing that the remaining "natural states of motions" were characterized not only by uniform speeds but also by rectilinear paths. In order to conceive of inertial motion, it is necessary to review the whole range of observable motions of material objects and imagine those motions if the effects of all known external influences were removed.

From this resulting set of ideal states of motion, it is necessary to identify the largest possible "equivalence class" of relatively uniform and rectilinear motions. These motions and configurations then constitute the basis for inertial measurements of space and time, i.e., inertial coordinate systems. Naturally inertial motions will then necessarily be uniform and rectilinear with respect to these coordinate systems, by definition.

Shortly thereafter (1644), Descartes presented the concept of inertial motion in his "Principles of Philosophy":

Each thing...continues always in the same state, and that which is once moved always continues to move...and never changes unless caused by an external agent... all motion is of itself in a straight line...every part of a body, left to itself, continues to move, never in a curved line, but only along a straight line.

Similarly, in Huygens' "The Motion of Colliding Bodies" (composed in the mid 1650's but not published until 1703), the first hypothesis was that

Any body already in motion will continue to move perpetually with the same speed in a straight line unless it is impeded.

Ultimately Newton incorporated this principle into his masterpiece, "Philosophiae Naturalis Principia Mathematica" as the first of his three "laws of motion"

- Every body continues in its state of rest, or of uniform motion in a right line, unless it is compelled to change that state by the forces impressed upon it.
- The change of motion is proportional to the motive force impressed, and is made in the direction of the right line in which that force is impressed.
- To every action there is always opposed an equal and opposite reaction; or, the mutual actions of two bodies upon each other are always equal, and directed to contrary parts.

These "laws" expresses the classical mechanical principle of relativity, asserting equivalence between the conditions of "rest" and "uniform motion in a right line". Since no distinction is made between the various possible *directions* of uniform motion, the principle also implies the equivalence of uniform motion in all directions in space.

Thus, if everything in the universe is a "body" in the sense of this law, and if we stipulate rules of force (such as Newton's second and third laws) that likewise do not distinguish between bodies at rest and bodies in uniform motion, then we arrive at a complete system of dynamics in which, as Newton said, "absolute rest cannot be determined from the positions of bodies in our regions". Corollary 5 of the Newton's Principia states

The motions of bodies included in a given space are the same among themselves, whether that space is at rest or moves uniformly forwards in a straight line without circular motion.

Of course, this presupposes that the words "uniformly" and "straight" have unambiguous meanings. Our concepts of uniform speed and straight paths are ultimately *derived* from observations of inertial motions, so the "laws of motion" are to some extent circular.

These laws were historically expressed in terms of inertial coordinate systems, which in turn are defined by the laws of motion. In other words, we define an inertial coordinate system as a system of space and time coordinates in terms of which inertia is homogeneous and isotropic, and then we announce the "laws of motion", which assert that inertia is homogeneous and isotropic with respect to inertial coordinate systems. Thus the "laws of motion" are true by definition. Their significance lies not in their truth, which is trivial, but in their applicability.

The empirical fact that there exist systems of inertial coordinates is what makes the concept significant. We have no *a priori* reason to expect that such coordinate systems exist, i.e., that the forces of nature would resolve themselves so coherently on this (or any other finite) basis, but they evidently do. In fact, it appears that not just one such coordinate system exists (which would be remarkable enough), but that infinitely many of them exist, in all possible states of relative motion.

To be precise, the principle of relativity asserts that for any material particle in any state of motion there exists an inertial coordinate system in terms of which the particle is (at least momentarily) at rest.

It's important to recognize that Newton's first law, by itself, is not sufficient to identify the systems of coordinates in terms of which all three laws of motion are satisfied. The first law serves to determine the shape of the coordinate axes and inertial paths, but it does not fully define a system of inertial coordinates, because the first law is satisfied in infinitely many systems of coordinates that are *not* inertial. The system of oblique xt coordinates illustrated below is an example of such a system.

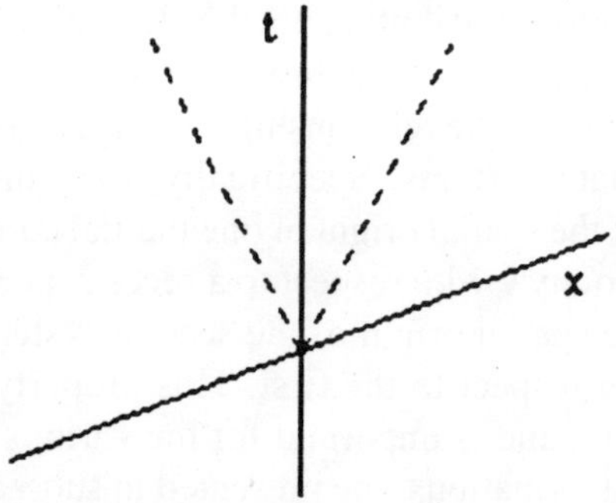

The two dashed lines indicate the paths of two identical objects, both initially at rest with respect to these coordinates and propelled outward from the origin by impulses forces of equal magnitude (acting against each other). Every object not subject to external forces moves with uniform speed in a straight line with respect to this coordinate system, so Newton's First Law of motion is satisfied, but the second law clearly is not, because the speeds imparted to these identical objects by equal forces are not equal. In other words, inertia is not isotropic with respect to these coordinates.

In order for Newton's Second Law to be satisfied, we not only need the coordinate axes to be straight and uniformly graduated relative to freely moving objects, we need the space axes to be aligned in time such that mechanical inertia is the same in all spatial directions (so that, for example, the objects whose paths are represented by the two dashed lines in the above figure have the same speeds).

This effectively establishes the planes of simultaneity of inertial coordinate systems. In an operational sense, Newton's Third Law is also involved in establishing the planes of simultaneity for an inertial coordinate system, because it is only by means of the Third Law that we can actually define "equal forces" as the forces necessary to impart equal "quantities of motion" (to use Newton's phrase). Of course, this doesn't imply that inertial coordinate systems are the "true" systems of reference. They are simply the most intuitive, convenient, and readily accessible systems, based on the inertial behaviour of material objects.

In addition to contributing to the definition of an inertial coordinate system, the third law also serves to establish a fundamental aspect of the relationships *between* relatively moving inertial coordinate systems. Specifically, the third law implies (requires) that if the spatial origin of one inertial coordinate system is moving at velocity v with respect to a second inertial coordinate system, then the spatial origin of the second system is moving at velocity –v with respect to the first. This property is sometimes called reciprocity, and is important for the various derivations of the Lorentz transformation to be presented in subsequent sections.

Based on the definition of an inertial coordinate system, and the isotropy of inertia with respect to such coordinates, it follows that two identical objects, initially at rest with respect to those coordinates and exerting a mutual force on each other, recoil by equal distances in equal times (in accord with Newton's third law). Assuming the lengths of stable material objects are independent of their spatial positions and orientations (spatial homogeneity and isotropy), it follows that we can synchronize distant clocks with identical particles ejected with equal forces from the mid-point between the clocks.

Of course, this operational definition of simultaneity is not new. It is precisely what Galileo described in his illustration of inertial motion onboard a moving ship. When he wrote that an object thrown with equal force will reach equal distances [in the same time], he was implicitly defining simultaneity at separate locations on the basis of inertial isotropy. This is crucial to understanding the significance of inertial coordinate systems.

The requirement for a particular object to be at rest with respect to the system suffices only to determine the direction of the "time axis", i.e., the loci of constant spatial position. Galileo and his successors realized (although they did not always explicitly state) that it is also necessary to specify the loci of constant temporal position, and this is achieved by choosing coordinates in such a way that mechanical inertia is isotropic.

Conceptually, to establish a complete system of space and time coordinates based on inertial isotropy, imagine that at each point in space there is an identically constructed cannon, and all these cannons are at rest with respect to each other. At one particular point, which we designate as the origin of our coordinates, is a clock and numerous identical cannons, each pointed at one of the other cannons out in space.

The cannons are fired from the origin, and when a cannonball passes one of the external cannons it triggers that external cannon to fire a reply back to the origin. Each cannonball has identifying marks so we can correlate each reply with the shot that triggered it, and with the identity of the replying cannon. The ith reply event is assigned the time coordinate $t_i = [t_{return}(i) - t_{send}(i)]/2$ seconds, and it is assigned space coordinates x_i, y_i, z_i based on the angular direction of the sending cannon and the radial distance $r_i = t_i$ cannon-seconds. This procedure would have been perfectly intelligible to Newton, and he would have agreed that it yields an inertial coordinate system, suitable for the application of his three laws of motion.

Naturally given one such system of coordinates, we can construct infinitely many others by simple spatial re-orientation of the space axes and/or translation of the spatial or temporal axes. All such transformations leave the speed of every object unchanged. An equivalence class of all such inertial coordinate systems is called an *inertial reference frame*.

For characterizing the mutual dynamical states of two material bodies, the associated inertial rest frames of the bodies are more meaningful than the mere distance between the bodies, because any inertial coordinate system possesses a fixed spatial *orientation* with respect to any other inertial coordinate system, enabling us to take account of tangential motion between bodies whose mutual distance is not changing. For this reason, the physically meaningful "relative velocity of two material bodies" is best defined as their reciprocal states of motion with respect to each others' associated inertial rest frame coordinates.

The principle of relativity does not tell us how two relatively moving systems of inertial coordinates are related to each other, but it does imply that this relationship can be determined empirically. We need only construct two relatively moving systems of inertial coordinates and compare them. Based on observations of coordinate systems with relatively low mutual speeds, and with the limited precision available at the time, Galileo and Newton surmised that if (x, t) is an inertial coordinate system then so is (x', t'), where

$$x' = x - vt \qquad t' = t$$

and v is the mutual speed between the origins of the two systems. This implies that relative speeds are simply additive. In other words, if a material object B is moving at the speed v in terms of inertial rest frame coordinates of A, and if an object C is moving in the same direction at the speed u in terms of inertial rest frame coordinates of B, then C is moving at the speed v + u in terms of inertial rest frame coordinates of A.

This conclusion may seem plausible, but it's important to realise that we are not free to arbitrarily adopt this or any other transformation and speed composition rule for the set of inertial coordinate systems, because those systems are already fully defined (up to insignificant scale factors) by the requirements for inertia to be homogeneous and isotropic and for momentum to be conserved. These properties suffice to determine the set of inertial coordinate systems *and* (therefore) the relationships between them. Given these conditions, the relationship between relatively moving inertial coordinate systems, whatever it may be, is a matter of empirical fact.

Of course, inertial isotropy is not the only possible basis for constructing spacetime coordinate systems. We could impose a different constraint to determine the loci of constant temporal position, such as a total temporal ordering of events. However, if we do this, we will find that mechanical inertia is generally not isotropic in terms of the resulting coordinate systems, so the usual symmetrical laws of mechanics will not be valid in terms of those coordinate systems (at least not if restricted to ponderable matter). Indeed this was the case for the ether theories developed in the late 19th century.

Such coordinate systems, while extremely awkward, would not be logically inconsistent. The choices we make to specify a coordinate system and to resolve spacetime intervals into separate spatial and temporal components are to some extent conventional, provided we are willing to disregard the manifest symmetry of physical phenomena. But since physics consists of identifying and understanding the symmetries of nature, the option of disregarding those symmetries does not appeal to most physicists.

By the end of the nineteenth century a new class of phenomena involving electric and magnetic fields had been incorporated into physics, and the concept of inertia was found to be applicable to these phenomena as well. For example, Maxwell's equations imply that a pulse of light conveys momentum. Hence the principle of inertia ought to apply to electromagnetism as well as to the motions of material bodies. In his 1905 paper "On the Electrodynamics of Moving Bodies" Einstein adopted this more comprehensive interpretation of inertia, basing the special theory of relativity on the proposition that

The laws by which the states of physical systems undergo changes are not affected, whether these changes of state be referred to the one or the other of two systems of [inertial] coordinates in uniform translatory motion.

This is nearly identical to Newton's Corollary 5. It's unfortunate that the word "inertial" was omitted, because, as noted above, uniform translatory motion is not sufficient to ensure that a system of coordinates is actually an inertial coordinate system. However, Einstein made it clear that he was indeed talking about inertial coordinate systems when he previously characterized them as coordinate systems "in which the equations of Newtonian mechanics hold good". Admittedly this is a somewhat awkward assertion in the context of Einstein's paper, because one of the main conclusions of the paper is that the equations of Newtonian mechanics do *not* precisely "hold good" with respect to inertial coordinate systems.

Recognizing this inconsistency, Sommerfeld added a footnote in subsequent published editions of Einstein's paper, qualifying the statement about Newtonian mechanics holding good "to the first approximation", but this footnote does not really clarify the situation.

Fundamentally, the class of coordinate systems that Einstein was trying to identify (the inertial coordinate systems) are those in terms of which inertia is homogeneous and isotropic, so that free objects move at constant speed in straight lines, and the force required to accelerate an object from rest to a given speed is the same in all directions. These conditions are just sufficient to determine a coordinate system in terms of which the symmetrical equations of mechanics hold good, but without pre-supposing the exact form of those equations.

Since light (i.e., an electromagnetic wave) carries momentum, and the procedure for constructing an inertial coordinate system described previously was based on the isotropy of momentum, it is reasonable to expect that pulses of light could be used in place of cannonballs, and we should arrive at essentially the same class of coordinate systems. In his 1905 paper this is how Einstein described the construction of inertial coordinate systems, implicitly asserting that the propagation of light is isotropic with respect to the same class of coordinate systems in terms of which mechanical inertia is isotropic.

In this respect it might seem as if he was treating light as a stream of inertial particles, and indeed his paper on special relativity was written just after the paper in which he introduced the concept of photons.

However, we know that light is not exactly like a stream of material particles, especially because we cannot conceive of light being at rest with respect to any system of inertial coordinates. The way in which light fits into the framework of inertial coordinate systems is considered.

We will find that although the principle of relativity continues to apply, and the definition of inertial coordinate systems remains unchanged, the relationship between relatively moving systems of inertial coordinate systems must be different than what Galileo and Newton surmised.

Chapter 2

The Relativity of Light

Light is arguably the phenomenon of nature with which we have the most conscious experience, by means of our sense of vision, and yet throughout most of human history very little seems to have been known about how vision works. Interestingly, from the very beginning there were at least two distinct concepts of light, existing side by side, as can be seen in some of the earliest known writings. For example, the description of creation in the biblical book of Genesis says light was created on the first day, and yet the sun, moon, and stars were not created until the fourth day "to give light upon the earth".

Evidently the word "light" is being used to signify two different things on the first and fourth days. For another example, Plato argued in Timaeus that there are two kinds of "fire" involved in our sense of vision, one coming from inside ourselves, emanating as visual rays from our eyes to make contact with distant objects, and another, which he called "daylight", that (when present) surrounds the visual rays from our eyes and facilitates the conveyance of the visual images.

These two kinds of "fire" correspond roughly with the later scholastic concepts of *lux* and *lumen*. The word lux was used to signify our visual sensations, whereas the word lumen referred to an external agent (such as light from the sun) that somehow participates in our sense of vision.

There was also, in ancient times, a competing theory of vision, according to which all objects naturally emit whole "images" (eidola) of themselves in small packets, and these enter our souls by way of our eyes. To account for our inability to see at night, it was thought that light from the sun or moon struck the objects and caused them to emit their images. This model of vision still entailed two distinct kinds of light: the facilitating illumination from the sun or moon, and the eidola emitted by ordinary objects.

This somewhat awkward conception of vision was improved by Ibn al-Haitham and later by Kepler, who argued that it is not necessary to assume whole objects emit multiple copies of themselves; we can simply consider each tiny part of an object as the source of rays emanating in all directions, and a sub-set of these rays intersecting in the eye can be re-assembled into an image of the object.

Until the end of the 17th century there was no evidence to indicate that rays of light propagated at a finite speed, and they were often assumed to be instantaneous. Only in 1689 with Roemer's observations of the moons of Jupiter, and even more convincingly in 1728 with Bradley's discovery of stellar aberration, did it become clear that the rays of lumen propagate through space with a characteristic finite speed. This suggested that light, and the energy it conveys, must have some mode of existence during the interval of time between its emission and its absorption.

Hence light became an entity or process in itself, rather than just a relation between entities, but again there were two competing notions as to the mode of existence. Two different analogies were conceived, based on the behaviour of ordinary material substances. Some thought light could be regarded as a stream of material corpuscles moving through empty space, whereas other believed light consists of undulations or waves in a pervasive material medium.

Each of these analogies was consistent with some of the attributes of light, but neither could be reconciled fully with all the attributes. For example, if light consists of material corpuscles, then according to Galilean relativity there should be an inertial reference frame with respect to which light is at rest in a vacuum, whereas in fact we never observe light in a vacuum to be at rest, nor even noticeably slow, with respect to any inertial reference frame.

On the other hand, if light is a wave propagating through a material medium, then the constituent parts of that medium should, according to Galilean relativity, behave inertially, and in particular should have a definite rest frame, whereas we find that light propagates best through regions (vacuum) in which there is no detectable material with a definite rest frame, and again we cannot conceive of light at rest in any inertial frame. Thus the behaviour of light defies realistic representation in terms of the behaviour of material substances within the framework of Galilean space and time, even if we consider just the classical attributes, let alone quantum phenomena.

By the end of the 19th century the inadequacy of both of the materialistic analogies for explaining the behaviour of light had become acute, because there was strong evidence that light exhibits two seemingly mutually exclusive properties. First, Maxwell showed how light can be regarded as a propagating electromagnetic wave, and as such the speed of propagation is obviously independent of the speed of the source.

Second, numerous experiments showed that light propagates at the same speed in all directions relative to the source, just as we would expect for streams of inertial corpuscles. Hence some of the attributes of light seemed to unequivocally support an emission theory, while others seemed just as unequivocally to support a wave theory. In retrospect it's clear that there was an underlying confusion regarding the terms of description, i.e., the systems of inertial coordinates, but this was far from clear at the time.

One of the first clues to unraveling the mystery was found in 1887, when Woldemar Voigt made a remarkable discovery concerning the ordinary wave equation. Recall that the wave equation for a time-dependent scalar field $\phi(x, t)$ in one dimension is

$$\frac{\partial^2 \phi}{\partial x^2} = \frac{1}{u^2} \frac{\partial^2 \phi}{\partial t^2}$$

where u is the propagation speed of the wave. This equation was first studied by Jean d'Alembert in the 18th century, and it applies to a wide range of physical phenomena. In fact it seems to represent a fundamental aspect of the relationship between space, time, and motion, transcending any particular application.

Traditionally it was considered to be valid only for a coordinate system x, t with respect to which the wave medium (presumed to be an inertial substance) is at rest and has isotropic properties, because if we apply a Galilean transformation to these coordinates, the wave equation is not satisfied with respect to the transformed coordinates. However, Galilean transformations are not the most general possible linear transformations. Voigt considered the question of whether there is *any* linear transformation that leaves the wave equation unchanged.

The general linear transformation between (X, T) and (x, t) is of the form

$$x = AX + BT \qquad t = CX + DT$$

for constants A, B, C, D. If we choose units of space and time so that the acoustic speed u equals 1, the wave equation in terms of (X, T) is simply $\partial^2\phi/\partial X^2 = \partial^2\phi/\partial T^2$.

To express this equation in terms of the transformed (*x*, *t*) coordinates, recall that the total differential of ϕ can be written in the form

$$d\phi = \frac{\partial\phi}{\partial x}dx + \frac{\partial\phi}{\partial t}dt$$

Also, at any constant T, the value of ϕ is purely a function of X, so we can divide through the above equation by dX to give

$$\frac{\partial\phi}{\partial X} = \left(\frac{\partial\phi}{\partial X}\right)_T = \frac{\partial\phi}{\partial x}\left(\frac{dx}{dX}\right)_T + \frac{\partial\phi}{\partial t}\left(\frac{dt}{dX}\right)_T$$

$$= A\frac{\partial\phi}{\partial x} + C\frac{\partial\phi}{\partial t}$$

Taking the partial derivative of this with respect to X then gives

$$\frac{\partial^2\phi}{\partial X^2} = A\frac{\partial^2\phi}{\partial X\partial x} + C\frac{\partial^2\phi}{\partial X\partial t}$$

Since partial differentiation is commutative, this can be written as

$$\frac{\partial^2\phi}{\partial X^2} = A\frac{\partial}{\partial x}\left(\frac{\partial\phi}{\partial X}\right) + C\frac{\partial}{\partial t}\left(\frac{\partial\phi}{\partial X}\right)$$

Substituting the prior expression for $\partial\phi/dX$ and carrying out the partial differentiations gives an expression for $\partial^2\phi/\partial X^2$ in terms of partials of ϕ with respect to x and t. Likewise we can derive an expression for $\partial^2\phi/\partial T^2$. Substituting into the wave equation gives

$$A^2 \frac{\partial^2\phi}{\partial x^2} + 2AC \frac{\partial^2\phi}{\partial x\partial t} + C^2 \frac{\partial^2\phi}{\partial t^2} = B^2 \frac{\partial^2\phi}{\partial x^2} + 2BD \frac{\partial^2\phi}{\partial x\partial t} + D^2 \frac{\partial^2\phi}{\partial t^2}$$

This is equivalent to the condition that $\phi(X, T)$ is a solution of the wave equation with respect to the X, T coordinates. Since the mixed partial generally varies along a path of constant second partial with respect to x or t, it follows that a necessary and sufficient condition for $\phi(x, t)$ to also be a solution of the wave equation in terms of the x, t coordinates is that the constants A, B, C, D of our linear transformation satisfy the relations

$$A^2 + C^2 = B^2 + D^2 \quad AC = BD$$

Furthermore, the differential of the space transformation is $dx = AdX + BdT$, so an increment with $dx = 0$ satisfies $dX/dT = -B/A$. This represents the velocity at which the spatial origin of the x, t coordinates is moving relative to the X, T coordinates. We will refer to this velocity as v. We also have the inverse transformation from (X, T) to (x, t):

$$X = \frac{D}{AD - BC^x} + \frac{-B}{AD - BC} t$$

$$T = \frac{-C}{AD - BC^x} + \frac{A}{AD - BC} t$$

Proceeding as before, the differential of this space transformation gives $dx/dt = B/D$ for the velocity of the spatial origin of the X, T coordinates with respect to the x, t coordinates, and this must equal $-v$. Therefore we have $B = -Av = -Dv$, and so $A = D$. It follows from the condition imposed by the wave equation that $B = C$, so both of these equal $-Av$. Our transformation can then be written in the form

$$x = A(X - vT) \quad x = A(T - vX)$$

The same analysis shows that the perpendicular coordinates y and z of the transformed system must be given by

$$y = A\sqrt{1-v^2}\,Y \quad z = A\sqrt{1-v^2}\,Z$$

In order to make the transformation formula for x agree with the Galilean transformation, Voigt chose A = 1, so he did not actually arrive at the Lorentz transformation, but nevertheless he had shown roughly how the wave equation could actually be relativistic – just like the dynamic behaviour of inertial particles – provided we are willing to consider a transformation of the space and time coordinates that differs from the Galilean transformation.

Had he considered the inverse transformation

$$X = \frac{1}{A(1-v^2)}(x + vt) \quad T = \frac{1}{A(1-v^2)}(t + vx)$$

he might have noticed that the determinant is $A^2(1-v^2)$, so to make this equal to 1 we must have $A = 1/(1-v^2)^{1/2}$, which not only implies y = Y and z = Z, but also makes the transformation formally identical to its inverse. In other words, he would have arrived at a completely relativistic framework for the wave equation. However, this was not Voigt's objective, and he evidently regarded the transformed coordinates x, y, z and t as merely a convenient parameterization for purposes of calculation, without attaching any greater significance to them.

Voigt's transformation was the first hint of how a wavelike phenomenon could be compatible with the principle of relativity, which is that there exist inertial coordinate systems in terms of which free motions are linear, inertia is isotropic, and every material object is instantaneously at rest with respect to one of these systems. None of this conflicts with the observed behaviour of light, because the motion of light is observed to be both linear and isotropic with respect to inertial coordinate systems. The fact that light is not at rest with respect to any system of inertial coordinates does not conflict with the principle of relativity if we agree that light is not a material object.

The incompatibility of light with the Galilean framework arises not from any conflict with the principle of relativity, but from the tacitly adopted empirical conclusion that two relatively moving systems of inertial coordinates are related to each other by Galilean transformations, so that the composition of co-linear speeds is simply additive. We aren't free to impose this assumption on the class of inertial coordinate systems, because they are fully determined by the requirement for inertia to be homogeneous and isotropic. There are no more adjustable parameters (aside from insignificant scale factors), so the composition of velocities with respect to relatively moving inertial coordinate systems is a matter to be determined empirically.

The basis of slowly moving reference frames, Galileo and Newton had inferred that the composition of speeds was simply additive. In other words, if a material object B is moving at the speed v in terms of inertial rest frame coordinates of a material object A, and if an object C is moving in the same direction at the speed u in terms of inertial rest frame coordinates of B, then Newton found that object C has the speed v + u in terms of the inertial rest frame coördinates of A. Toward the end of the nineteenth century, more precise observations revealed that is not quite correct. It was found that the speed of object C in terms of inertial rest frame coordinates of A is not v + u, but rather $(v + u)/(1 + uv/c^2)$, where c is the speed of light in a vacuum.

Obviously these conclusions would be identical if the speed of light was infinitely great, which was still considered a real possibility in Galileo's day. Many people, including Descartes, regarded rays of light as instantaneous. Even Newton's Opticks, published in 1704, made allowances for the possibility that "light be propagated in an instant" (although Newton himself was persuaded by Roemer's observations that light has a finite speed).

Hence it can be argued that the principles of Galileo and Einstein are essentially identical in both form and content. The only difference is that Galileo assessed the propagation of light to be "if not instantaneous then extraordinarily fast", and thus could neglect the term uv/c^2, especially since he restricted his considerations to the movements of material objects, whereas subsequently it became clear that the speed of light has a finite value, and it was necessary to take account of the uv/c^2 term when attempting to incorporating the motions of light and high-speed particles into the framework of mechanics.

The empirical correspondence between inertial isotropy and lightspeed isotropy can be illustrated by a simple experiment.

Three objects, A, B, and C, at rest with respect to each other can be arranged so that one of them is at the midpoint between the other two (the midpoint having been determined using standard measuring rods at rest with respect to those objects). The two outer objects, A and C, are equipped with identical clocks, and the central object, B, is equipped with two identical cannons. Let the two cannons in the centre be fired simultaneously in opposite directions toward the two outer objects, and then at a subsequent time let object B emit a flash of light.

If the arrivals of the cannonball and light coincide at A, then they also coincide at C, signifying that the propagation of light is isotropic with respect to the same system of coordinates in terms of which mechanical inertia is isotropic, as illustrated in the figure below.

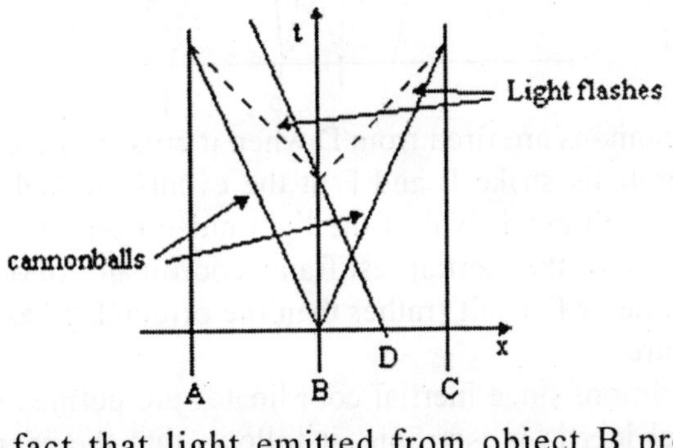

The fact that light emitted from object B propagates isotropically with respect to B's inertial rest frame might seem to suggest that light can be treated as an inertial object within the Galilean framework, just like cannon-balls. However, we also find that if the light is emitted at the same time and place from an object D that is moving with respect to B, the light's speed is still isotropic with respect to B's inertial rest frame.

Now, this might seem to suggest that light is a disturbance in a material medium in which the objects A, B, C just happen to be at rest, but this is ruled out by the fact that it applies regardless of the state of (uniform) motion of those objects. Naturally this implies that the flash of light propagates isotropically with respect to the inertial rest coordinates of object D as well. To demonstrate this, we could arrange for two other bodies, denoted by E and F, to be moving at the same speed as D, and located an equal distance from D in opposite directions. Then we could fire two identically constructed cannons (at rest with respect to D) in opposite directions, toward E and F. The results are illustrated below.

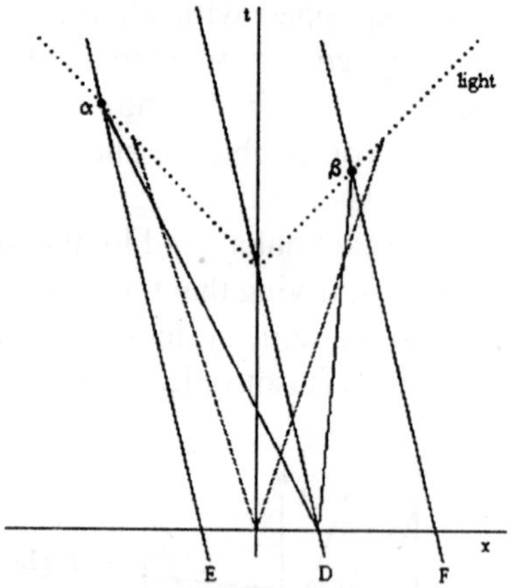

The cannons are fired from D when it crosses the x axis, and the cannon-balls strike E and F at the events marked α and β, coincident with the arrival of the light pulse from D. Obviously the time axis for the inertial rest frame coordinates of object D is the worldline of D itself (rather than the original "t" axis shown on the figure).

In addition, since inertial coordinates are defined such that mechanical inertia is isotropic, it follows that the cannon-balls fired from identical cannons at rest with D are moving with equal and opposite speeds with respect to D's inertial rest coordinates, and since E and F are at equal distances from D, it also follows that the events a and b are simultaneous with respect to the inertial rest coordinates of D. Hence, not only is the time axis of D's rest frame slanted with respect to B's time axis, the spatial axis of D's rest frame is equally slanted with respect to B's spatial axis.

Several other important conclusions can be deduced from this figure. For example, with respect to the original x, t coordinate system, the speeds of the cannon-balls from D are not given by simply adding (or subtracting) the speed of the cannon-balls with respect to D's rest frame to (or from) the speed of D with respect to the x, t coordinates. Since momentum is explicitly conserved, this implies that the inertia of a body increases with it's velocity (i.e., kinetic energy).

We should also note that although the *speed* of light is isotropic with respect to any inertial spacetime coordinates, independent of the motion of the source, it is not correct to say that the light itself is isotropic. The relationship between the frequency (and energy) of the light with respect to the rest frame of the emitting body and the frequency (and energy) of the light with respect to the rest frame of the receiving body *does* depend on the relative velocity between those two massive bodies.

Incidentally, notice that we can rule out the possibility of object B and D dragging the light medium along with them, because they are moving through the *same* region of space at the same time, and they can't *both* be dragging the same medium in opposite directions. This is in contrast to the case of (for example) acoustic pressure waves in a material substance, because in that case a recognizable material substance determines the unique isotropic frame, whereas in the case of light we're unable to identify any definite material medium, so the medium has no definite rest frame.

The first person to discern the true relationship between relatively moving systems of inertial coordinate systems was Hendrik Antoon Lorentz. Not surprisingly, he arrived at this conception in a rather indirect and laborious way, and didn't immediately recognize that the class of coordinate systems he had discovered (and which he called "local coordinate" systems) were none other than Galileo's inertial coordinate systems.

Incidentally, although Lorentz and Voigt knew and corresponded with each other, Lorentz apparently was not aware of Voigt's earlier work on coordinate transformations that leave the wave equation invariant, and so that work had no influence on Lorentz's search for coordinate systems in terms of which Maxwell's equations are invariant. Unlike Voigt, Lorentz derived the transformation in two separate stages. He first developed the "local time" coordinate, and only years later came to the conclusion (after, but independently of, Fitzgerald) that a "contraction" of spatial length was also necessary in order to account for the absence of second-order effects in Michelson's experiment.

Lorentz began with the absolute ether frame coordinates t and x, in terms of which every event can be assigned a unique space-time position (t, x), and then he considered a system moving with the velocity v in the positive x direction. He applied the traditional Galilean transformation to assign a new set of coordinates to every event.

Thus an event with ether-frame coordinates t, x is assigned the new coordinates $x'' = x - vt$ and $t'' = t$. Then he tentatively proposed an additional transformation that must be applied to x″, t″ in order to give coordinates in terms of which Maxwell's equations apply in their standard form.

Lorentz was not entirely clear about the physical significance of these "local" coordinates, but it turns out that *all* physical phenomena conform to the same isotropic laws of physics when described in terms of these coordinates. (Lorentz's notation made use of the parameter $\beta = 1/\gamma = 1/(1-v^2)^{1/2}$ and another constant which he later determines to be 1). Taking units such that c = 1, his equations for the local coordinates x′ and t′ in terms of the Galilean coordinates which we are calling x′ and t″ are

$$x' = \frac{x''}{\gamma} \qquad t' = \gamma t'' - \frac{vx''}{\gamma}$$

Recall that the traditional Galilean transformation is $x'' = x - vt$ and $t'' = t$, so we can make these substitutions to give the complete transformation from the original ether rest frame coordinates x, t to the local coordinates moving with speed v

$$x' = \frac{x - vt}{\gamma} \qquad t' = \frac{t - vx}{\gamma}$$

These effective coordinates enabled Lorentz to explain how two relatively moving observers, each using his own local system of coordinates, both *seem* to remain at the centre of expanding spherical light waves originating at their point of intersection, as illustrated below.

The x and x″ axes represent the respective spatial coordinates (say, in the east/west direction), and the t and t′ axes represent the respective time coordinates. One observer is moving through time along the t axis, and the other has some relative westward velocity as he moves through time along the t′ axis. The two observers intersect at the event labeled O, where they each emit a pulse of light.

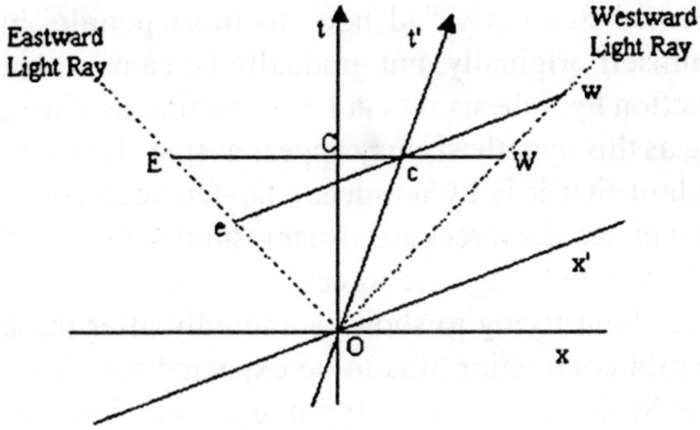

Those light pulses emanate away from O along the dotted lines. Subsequently the observer moving along the t axis finds himself at C, and according to his measures of space and time the outward going light waves are at E and W at that same instant, which places him at the midpoint between them. On the other hand, the observer moving along t′ axis finds himself at point c, and according to his measures of space and time the outward going light waves are at e and w at this instant, which implies that *he* is at the midpoint between them.

Thus Lorentz discovered that by means of the "fictitious" coordinates x′, t′ it was possible to conceive of a class of relatively moving coordinate systems with respect to which the speed of light is invariant. He went beyond Voigt in the realization that the existence of this class of coordinate systems ensures the appearance of relativity, at least for optical phenomena, and yet, like Voigt, he still tended to regard the "local coordinates" as artificial.

Having been derived specifically for electromagnetism, it was not clear that the same transformations should apply to all physical phenomena, including inertia, gravity, and whatever forces are responsible for the stability of matter – at least not without simply hypothesizing this to be the case.

However, Lorentz was dissatisfied with the proliferation of hypotheses that he had made in order to arrive at this theory. The same criticism was made in a contemporary review of Lorentz's work by Poincare, who chided him with the remark "hypotheses are what we lack least".

The most glaring of these was the hypothesis of contraction, which seemed distinctly "ad hoc" to most people, including Lorentz himself originally, but gradually he came to realise that the contraction hypothesis was not as unnatural as it might seem. Surprising as this hypothesis may appear at first sight, yet we shall have to admit that it is by no means far-fetched, as soon as we assume that molecular forces are also transmitted through the ether, like the electric and magnetic forces...

He set about trying to show (admittedly after the fact) that the Fitzgerald contraction was to be expected based on what he called the Molecular Force Hypothesis and his theorem of Corresponding States.

CORRESPONDING STATES

In 1889 Oliver Heaviside deduced from Maxwell's equations that the electric and magnetic fields on a spherical surface of radius r surrounding a uniformly moving electric charge e are radial and circumferential respectively, with magnitudes

$$|E| = \frac{e}{r^2}\frac{1-v^2}{\left(1-v^2\sin(\theta)^2\right)^{3/2}} \qquad |B| = |E|\, v\sin(\theta)$$

where θ is the angle relative to the direction of motion with respect to the stationary frame of reference. (We have set $c = 1$ for clarity). The left hand equation implies that, in comparison with a stationary charge, the electric field strength at a distance r from a moving charge is less by a factor of $1-v^2$ in the direction of motion, and greater by a factor of $1/(1-v^2)^{1/2}$ in the perpendicular directions. Thus the strength of the electric field of a moving charge is anisotropic. These equations imply that

$$E^2 - B^2 = \left(\frac{e}{r^2}\right)^2\left(\frac{1-v^2}{1-v^2\sin(\theta)^2}\right)$$

which Heaviside recognized as the *convection potential*, i.e., the scalar field whose gradient is the total electromagnetic force on a co-moving charge at that relative position. This scalar is invariant under Lorentz transformations, and it follows from the above formula that the cross-section of surfaces of constant potential are described by

$$r^2 = \left(\frac{e}{\sqrt{E^2 - B^2}}\right)\left(\frac{1 - v^2}{1 - v^2 \sin(\theta)^2}\right)$$

This is the equation of an ellipse, so Heaviside's formulas imply that the surfaces of constant potential are ellipsoids, shortened in the direction of motion by the factor $(1-v^2)^{1/2}$. From the modern perspective the contraction of characteristic lengths in the direction of motion is an immediate corollary of the fact that Maxwell's equations are Lorentz covariant, but at the time the idea of anisotropic changes in length due to motion was regarded as a distinct and somewhat unexpected attribute of electromagnetic fields.

It wasn't until 1896 that Searle explicitly pointed out that Heaviside's formulas imply the contraction of surfaces of constant potential into ellipsoids, but already in 1889 it seems that Heaviside's findings had prompted an interesting speculation as to the deformation of stable material objects in uniform motion.

George Fitzgerald corresponded with Heaviside, and learned of the anisotropic variations in field strengths for a moving charge, and this was at the very time when he was struggling to understand the null result of the latest Michelson and Morley ether drift experiment (performed in 1887). It occurred to Fitzgerald that the null result would be explained if the material comprising Michelson's apparatus contracts in the direction of motion by the factor $(1-v^2)^{1/2}$, and moreover that this contraction was not entirely implausible, because, as he wrote in a brief letter to the American journal *Science* in 1889

We know that electric forces are affected by the motion of the electrified bodies relative to the ether and it seems a not improbable supposition that the molecular forces are affected by the motion and that the size of the body alters consequently.

A few years later (1892) Lorentz independently came to the same conclusion, and proceeded to explain in detail how the variations in the electromagnetic field implied by Maxwell's equations actually result in a proportional contraction of matter – at least if we assume the forces responsible for the stability of matter are affected by motion in the same way as the forces of electromagnetism.

This latter assumption Lorentz called the "molecular force hypothesis", admitting that he had no real justification for it (other than the fact that it accounted for Michelson's null result). On the basis of this hypothesis, Lorentz showed that the description of the equilibrium configuration of a uniformly moving material object in terms of its "local coordinates" is identical to the description of the same object at absolute rest in terms of the ether rest frame coordinates. He called this the *theorem of corresponding states*.

To illustrate, consider a small bound spherical configuration of matter at rest in the ether. We assume the forces responsible for maintaining the spherical structure of this particle are affected by uniform motion through the ether in exactly the same way as are electromagnetic forces, which is to say, they are covariant with respect to Lorentz transformations.

These forces may propagate at any speed (at or below the speed of light), but it is most convenient for descriptive purposes to consider forces that propagate at precisely the speed of light (in terms of the fixed rest frame coordinates of the ether), because this automatically ensures Lorentz covariance. A wave emanating from the geometric centre of the particle at the speed c would expand spherically until reaching the radius of the configuration, where we can imagine that it is reflected and then contracts spherically back to a point (like a spatial filter) and re-expands on the next cycle. This is illustrated by the left-hand cycle below.

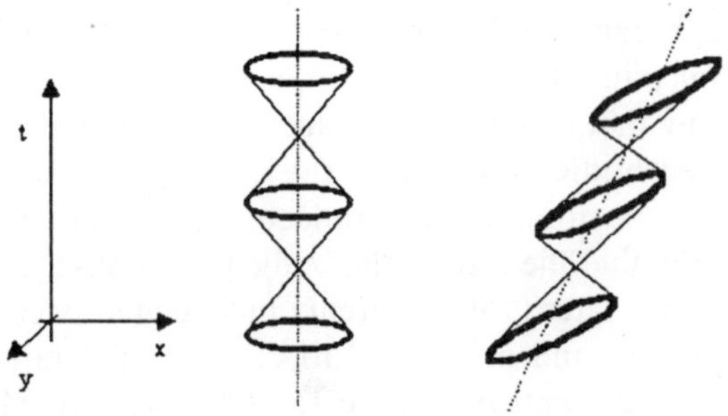

Only two spatial dimensions are shown in this figure. (In four-dimensional spacetime each shell is actually a sphere). Now, if we consider an intrinsically identical configuration of matter in uniform motion relative to the putative rest frame of the ether, and if the equilibrium shape is maintained by forces that are Lorentz covariant, just as is the propagation of electromagnetic waves, then it must still be the case that an electromagnetic wave can expand from the centre of the configuration to the perimeter, and be reflected back to the centre in a coherent pattern, just as for the stationary configuration. This implies that the absolute shape of the configuration must change from a sphere to an ellipsoid, as illustrated by the right-hand figure above.

The spatial size of the particle in terms of the ether rest frame coordinates is just the intersection of a horizontal time slice with the region swept out by the perimeter of the configuration. For any given characteristic particle, since there is no motion relative to the ether in the transverse direction, the size in the transverse direction must be unaffected by the motion. Thus the widths of the configurations in the "y" direction in the above figures are equal. The figure below shows more detailed side and top views of one cycle of a stationary and a moving particle (with motions referenced to the rest frame of the putative ether).

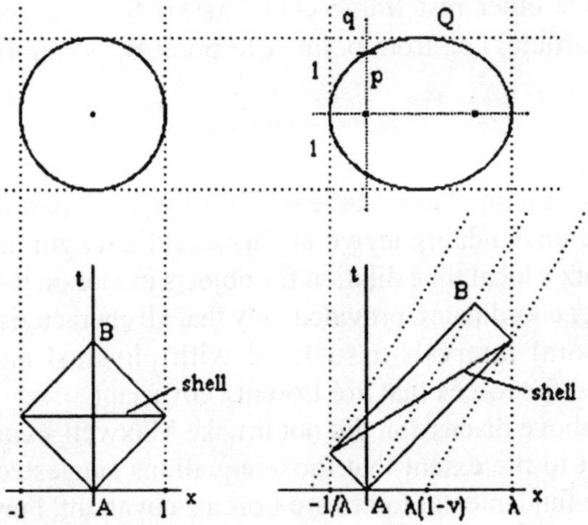

It's understood that these represent corresponding states, i.e., intrinsically identical equilibrium configurations of matter, whose spatial shapes are maintained by Lorentz covariant forces. In each case the geometric centre of the configuration progresses from point A to point B in the respective figure.

The right-hand configuration is moving with a speed v in the positive x direction. It can be shown that the transverse sizes of the configurations are equal if the projected areas of the cross-sectional side views (the lower figures) are equal. Thus, light emanating from point A of the moving particle extends a distance $1/\lambda$ to the left and a distance λ to the right, where λ is a constant function of v. Specifically, we must have

$$\lambda = \sqrt{\frac{1+v}{1-v}}$$

where we have set c = 1 for clarity. The leading edge of the shaft swept out by the moving shell crosses the x axis at a distance $\lambda(1-v)$ from the centre point A, which implies that the object's instantaneous spatial extent from the centre to the leading edge is only

$$(1-v) = \sqrt{\frac{1+v}{1-v}} = \sqrt{1-v^2}$$

Likewise it's easy to see that the elapsed time (according to the putative ether rest frame coordinates) for one cycle of the moving particle, i.e., from point A to point B, is simply

$$\lambda + \frac{1}{\lambda} = \frac{2}{\sqrt{1-v^2}}$$

compared with an elapsed time of 2 for the same particle at rest. Hence we unavoidably arrive at Fitzgerald's length contraction and Lorentz's local time dilation for objects in motion with respect to the x, y, t coordinates, provided only that all characteristic spatial and temporal intervals associated with physical entities are maintained for forces that are Lorentz covariant.

The above discussion did not invoke Maxwell's equations at all, except to the extent that those equations suggested the idea that all the fundamental forces are Lorentz covariant. Furthermore, we have so far omitted consideration of one very important force, namely, the force of *inertia*. We assumed the equilibrium configurations of matter were maintained by certain forces, but if we consider oscillating configurations, we see that the periodic shapes of such configurations depend not only on the binding force(s) but also on the inertia of the particles.

Therefore, in order to arrive at a fully coherent theorem of corresponding states, we must assume that inertia itself is Lorentz covariant. As Lorentz wrote in his 1904 paper…the proper relation between the forces and the accelerations will exist… if we suppose that the masses of all particles are influenced by a translation to the same degree as the electromagnetic masses of the electrons.

In other words, we must assume the inertial mass (resistance to acceleration) of every particle is Lorentz covariant, which implies that the mass has transverse and longitudinal components that vary in a specific way when the particle is in motion. Now, it was known that some portion of a charged object's resistance to acceleration is due to self-induction, because a moving charge constitutes an electric current, which produces a magnetic field, which resists changes in the current.

Not surprisingly, this resistance to acceleration is Lorentz covariant, because it is a purely electromagnetic effect. At one time it was thought that perhaps *all* mass (even of electrically neutral particles) might be electromagnetic in origin, and some even hoped that gravity and the unknown forces governing the stability of matter would also someday be shown to be electromagnetic, leading to a totally electromagnetic world view. (Ironically, at this same time, others were trying to maintain the mechanical world view, by seeking to explain the phenomena of electromagnetism in terms of mechanical models).

If in fact all physical effects are ultimately electromagnetic, one could plausibly argue that Lorentz had succeeded in developing a constructive account of relativity, based on the known properties of electromagnetism. Essentially this would have resolved the apparent conflict between the Galilean relativity of mechanics and Lorentzian relativity of electromagnetism, by asserting that there is no such thing as mechanics, there is only electromagnetism.

Then, since electromagnetism is Lorentz covariant, it would follow that everything is Lorentz covariant. However, it was already known (though perhaps not well known) when Lorentz wrote his paper in 1904 that the electromagnetic world view is not tenable. Poincare pointed this out in his 1905 Palermo paper, in which he showed that the assumption of a purely electromagnetic electron was self-consistent only with the degenerate solution of no charge density at all. Essentially, the linearity of Maxwell's equations implies that they can not possibly yield stable bound configurations of charge. Poincare wrote

We must then admit that, in addition to electromagnetic forces, there are also non-electromagnetic forces or bonds. Therefore, we need to identify the conditions that these forces or bonds must satisfy for electron equilibrium to be undisturbed by the [Lorentz] transformation.

In the remainder of this remarkable paper, Poincare derives general conditions that Lorentz covariant forces must satisfy, and considers in particular the force of gravity. The most significant point is that Poincare had recognized that Lorentz had reached the limit of his constructive approach, and instead he (Poincare) was proceeding not to deduce the necessity of relativity from the phenomena of electromagnetism or gravity, but rather to deduce the necessary attributes of electromagnetism and gravity from the principle of relativity. In this sense it is fair to say that Poincare originated a theory of relativity in 1905 (simultaneously with Einstein).

On the other hand, both Poincare and Lorentz continued to espouse the view that relativity was only an apparent fact, resulting from the circumstance that our measuring instruments are necessarily affected by absolute motion in the same way as are the things being measured. Thus they believed that the speed of light was actually isotropic only with respect to one single inertial frame of reference, and it merely *appeared* to be isotropic with respect to all the others.

Of course, Poincare realized full well (and indeed was the first to point out) that the Lorentz transformations form a group, and the symmetry of this group makes it impossible, even in principle, to single out one particular frame of reference as the true absolute frame (in which light actually does propagate isotropically). Nevertheless, he and Lorentz both argued that there was value in maintaining the belief in a true absolute rest frame, and this point of view has continued to find adherents down to the present day.

As a historical aside, Oliver Lodge claimed that Fitzgerald originally suggested the deformation of bodies as an explanation of Michelson's null result while sitting in my study at Liverpool and discussing the matter with me. The suggestion bore the impress of truth from the first. Interestingly, Lodge interpreted Fitzgerald as saying not that objects contract in the direction of motion but that they expand in the transverse direction. We saw in the previous section how Voigt's derivation of the Lorentz transformation left the scale factor undetermined, and the evaluation of this factor occupied a surprisingly large place in the later writings of Lorentz, Poincare, and Einstein. In his book *The Ether of Space* (1909) Lodge provided an explanation for why he believed the effect of motion should be a transverse expansion rather than a longitudinal contraction. He wrote

When a block of matter is moving through the ether of space its cohesive forces across the line of motion are diminished, and consequently in that direction it expands.

Lodge's reliability is suspect, since he presents this as an "explanation" not only of Fitzgerald's suggestion but also of Lorentz's theory, which it definitely is not. But more importantly, Lodge's misunderstanding highlights one of the drawbacks of conceiving of the deformation effect as arising from variations in electromagnetic forces. In order to give a coherent account of phenomena, the *lengths* of objects must vary in exactly the same proportion as the *distances* between objects.

It would be quite strange to suppose that the transverse distances between (neutral and widely separated) objects would increase by virtue of being set in motion along parallel lines. In fact, it is not clear what this would even mean. If three or more objects were set in parallel motion, in which direction would they be deflected? And what could be the cause of such a deflection? Neutral objects at rest exert a small attractive force on each other (due to gravity), but diminishing this net force of cohesion would obviously not cause the objects to repel each other. Oddly enough, if Lodge had focused on the temporal instead of the spatial effects of motion, his reasoning would have approximated a valid justification for time dilation. This justification is often illustrated in terms two mirror in parallel motion, with a pulse of light bouncing between them. In this case the motion of the mirrors actually does diminish the frequency of bounces, relative to the stationary ether frame, because the light must travel further between each reflection.

Thus the time intervals "expand" (i.e., dilate). Given this time dilation of the local moving coordinates, it's fairly obvious that there must be a corresponding change in the effective space coordinate (since spatial lengths are directly related to time intervals by $dx = vdt$). In other words, if an observer moves at speed v relative to the ground, and passes over an object of length L at rest on the ground, the length of the object as assessed by the moving observer is affected by his measure of time.

Since he is moving at speed v, the length of the object is vdt, where dt is the time it takes him to traverse the length of the object – but *which* "dt" will he use?

Naturally if he bases his length estimate on the measure of the time interval recorded on a ground clock, he will have $dt = L/v$, so he will judge the object to be $v(L/v) = L$ units in length. However, if he uses his own effective time as indicated on his own co-moving transverse light clock, he will have $dt' = dt\,(1-v^2)^{1/2}$, so the effective length is $v[(L/v)(1-v^2)^{1/2}] = L(1-v^2)^{1/2}$. Thus, effective length contraction (and *no* transverse expansion) is logically unavoidable given the effective time dilation.

It might be argued that we glossed over an ambiguity in the above argument by considering only light clocks with pulses moving transversely to the motion of the mirrors, giving the relation $dt' = dt(1-v^2)^{1/2}$. If, instead, we align the axis between the mirrors with the direction of travel, we get $dt' = dt(1-v^2)$, so it might seem we have an ambiguous measure of local time, and therefore an ambiguous prediction of length contraction since, by the reasoning given above, we would conclude that an object of rest-length L has the effective length $L(1-v^2)$.

However, this fails to account for the contraction of the longitudinal distance between the mirrors (when they are arranged along the axis of motion). Since by construction the speed of light is c in terms of the local coordinates for the clock, the very same analysis that implies length contraction for objects moving relative to the ether rest frame coordinates also implies the same contraction for objects moving relative to the new local coordinates. Thus the clock is contracted in the longitudinal direction relative to the ground's coordinates by the same factor that objects on the ground are contracted in terms of the moving coordinates.

The amount of spatial contraction depends on the amount of time dilation, which depends on the amount of spatial contraction, so it might seem as if the situation is indeterminate. However, all but one of the possible combinations are logically inconsistent. For example, if we decided that the clock was shortened by the full longitudinal factor of $(1-v^2)$, then there would be *no* time dilation at all, but with no time dilation there would be no length contraction, so this is self-contradictory.

The only self-consistent arrangement that reconciles each reference frame's local measures of longitudinal time and length is with the factor $(1-v^2)^{1/2}$ applied to both. This also agrees with the transverse time dilation, so we have isotropic clocks with respect to the local (i.e., inertial) coordinates of any uniformly moving frame, and by construction the speed of light is c with respect to each of these systems of coordinates. This is illustrated by the figures below, showing how the spacetime pattern of reflecting light rays imposes a skew in both the time and the space axes of relatively moving systems of coordinates.

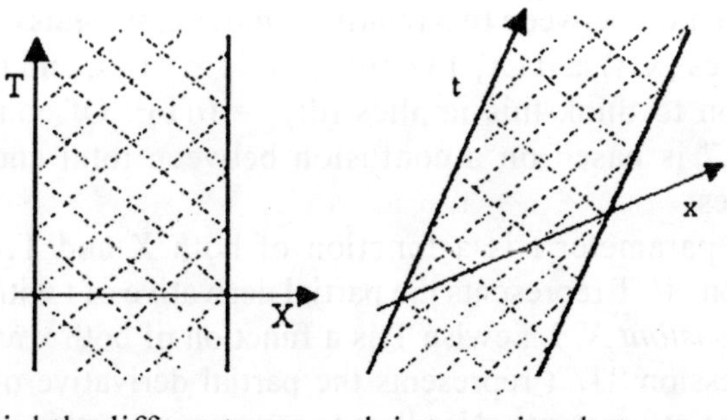

A slightly different approach is to notice that, according to a "transverse" light clock, we have the partial derivative "t/"T = 1/ $(1-v^2)^{1/2}$ along the absolute time axis, i.e., the line X = 0.

Integrating gives $t = (T - f(X))/(1-v^2)^{1/2}$ where f(x) is an arbitrary function of X. The question is: D*oes there exist* a function f(X) that will yield physical relativity?

If such a function exists, then obviously the resulting coordinates are the ones that will be adopted as the rest frame by any observer at rest with respect to them. Such a function does indeed exist, namely, f(X) = vX, which gives $t = (T-vX)/(1-v^2)^{1/2}$. To show reciprocity, note that X = vT along the t axis, so we have $t = T(1-v^2)/(1-v^2)^{1/2}$, which gives $T = t/(1-v^2)^{1/2}$ and so “T/”t = $1/(1-v^2)^{1/2}$.

As we've seen, this same transformation yields relativity in the longitudinal direction as well, so there does indeed exist, for any object in any state of motion, a coordinate system with respect to which all optical phenomena are isotropic, and as a matter of empirical fact this is precisely the same class of systems invoked by Galileo's principle of mechanical relativity, the inertial systems, i.e., coordinate systems with respect to which mechanical inertia is isotropic.

Lorentz noted that the complete reciprocity and symmetry between the “true” rest frame coordinates and each of the local effective coordinate systems may seem surprising at first. As he said in his Leiden lectures in 1910. The behaviour of measuring rods and clocks in translational motion, when viewed superficially, gives rise to a *remarkable paradox*, which on closer examination, however, vanishes.

The seeming paradox arises because the Lorentz transformation between two relatively moving systems of inertial coordinates (x, t) and (X, T) implies “t/”T = “T/”t, and there is a temptation to think this implies $(dt)^2 = (dT)^2$. Of course, this “paradox” is based on a confusion between total and partial derivatives.

The parameter t is a function of both X and T, and the expression “t/”T represents the partial derivative of t with respect to T *at constant X*. Likewise T is a function of both x and t, and the expression “T/”t represents the partial derivative of T with respect to t *at constant x*. Needless to say, there is nothing logically inconsistent about a transformation between (x, t) and (X, T) such that $(\partial t/\partial T)_X$ equals $(\partial T/\partial t)_x$, so the “paradox” vanishes.

The writings of Lorentz and Poincare by 1905 can be assembled into a theory of relativity that is operationally equivalent to the modern theory of special relativity, although lacking the conceptual clarity and coherence of the modern theory.

Lorentz was justifiably proud of his success in developing a theory of electrodynamics that accounted for all the known phenomena, explaining the apparent relativity of these phenomena, but he was also honest enough to acknowledge that the success of his programme relied on unjustified hypotheses, the most significant of which was the hypothesis that inertial mass is Lorentz covariant.

To place Lorentz's achievement in context, recall that toward the end of the 19th century it appeared electromagnetism was not relativistic, because the property of being relativistic was equated with being invariant under Galilean transformations, and it was known that Maxwell's equations (unlike Newton's laws of mechanics) do not possess this invariance.

Lorentz, prompted by experimental results, discovered that Maxwell's equations actually are relativistic, in the sense of his theorem of corresponding states, meaning that there are relatively moving coordinate systems in terms of which Maxwell's equations are still valid.

But these systems are not related by Galilean transformations, so it still appeared that mechanics (presumed to be Galilean covariant) and electrodynamics were not mutually relativistic, which meant it ought to be possible to discern second-order effects of absolute motion by exploiting the difference between the Galilean covariant of mechanics and Lorentz covariance of electromagnetism.

However, all experiments refuted this expectation. In other words, it was found empirically that electromagnetism and mechanics are mutually relativistic (at least to second order). Hence the only possible conclusion is that either the known laws of electromagnetism or the known laws of mechanics must be subtly wrong.

Either the correct laws of electromagnetism must really be Galilean covariant, or else the correct laws of inertial mechanics must really be Lorentz covariant.

At this point, in order to "save the phenomena", Lorentz simply assumed that inertial mass is Lorentz covariant. Of course, he had before him the example of self-induction of charged objects, leading to the concept of electromagnetic mass, which is manifestly Lorentz covariant, but, as Poincare observed, it is not possible (and doesn't even make sense) for the intrinsic mass of elementary particles to be electromagnetic in origin.

Hence the hypothesis of Lorentz covariance for inertia (and therefore inertial mechanics) is not a "constructive" deduction; it is not even implied by the molecular force hypothesis (because there is no reason to suppose that anything analogous to "self-induction" of the unknown molecular forces is ultimately responsible for inertia); it is simply a hypothesis, motivated by empirical facts. This does not diminish Lorentz's achievement, but it does undercut his comment that "Einstein simply postulates what we have deduced... from the fundamental equations of the electromagnetic field".

In saying this, Lorentz overlooked that fact that the Lorentz covariance of mechanical inertia cannot be deduced from the equations of electromagnetism. He simply postulated it, no less than Einstein did.

Much of the confusion over whether Lorentz deduced or postulated his results is due to confusion between the two aspects of the problem. First, it was necessary to determine that Maxwell's equations are Lorentz covariant. This was in fact deduced by Lorentz from the laws themselves, consistent with his claim. But in order to arrive at a complete theory of relativity (and in particular to account for the second-order null results) it is also necessary to determine that mechanical inertia (and molecular forces, and gravity) are all Lorentz covariant.

This proposition was not deduced by Lorentz (or anyone else) from the laws of electromagnetism, nor could it be, because it does not follow from those laws. It is merely postulated, just as we postulate the conservation of energy, as an organizing principle, justified by it's logical cogency and empirical success. As Poincare clearly explained in his Palermo paper, the principle of relativity itself emerges as the only reliable guide, and this is as true for Lorentz's approach as it is for Einstein's, the main difference being that Einstein recognized this principle was not only necessary, but also that it obviated the detailed assumptions as to the structure of matter. Hence, even with regard to electromagnetism (let alone mechanics) Lorentz could write in the 1915 edition of his Theory of Electrons that:

I should certainly have given a more prominent place to Einstein's theory of relativity, by which the theory of electromagnetic phenomena in moving systems gains a simplicity that I had not been able to attain.

Nevertheless, as mentioned previously, Lorentz and Poincare both continued to espouse the merits of the absolute interpretation of relativity, although Poincare's seemed to regard the distinction as merely conventional. For example, in a 1912 lecture he said the new conception according to which space and time are no longer two separate entities, but two parts of the same whole, which are so intimately bound together that they cannot be easily separated is a new convention [that some physicists have adopted] Not that they are constrained to do so; they feel that this new convention is more comfortable, that's all; and those who do not share their opinion may legitimately retain the old one, to avoid disturbing their ancient habits. Between ourselves, let me say that I feel they will continue to do so for a long time still.

Sadly, Poincare died just two months later, but his prediction has held true, because to this day the "ancient habits" regarding absolute space and time persist. There are today scientists and philosophers who argue in favour of what they see as Lorentz's constructive approach, especially as a way of *explaining* the appearance of relativity, rather than merely accepting relativity in the same way we accept (for example) the principle of energy conservation.

However, as noted above, the constructiveness of Lorentz's approach begins and ends with electromagnetism, the rest being conjecture and hypothesis, so this argument in favour of the Lorentzian view is misguided. But setting this aside, is there any merit in the idea that the absolutist approach effectively *explains* the appearance of relativity?

To answer this question, we must first clearly understand what precisely is to be explained when one seeks to "explain" relativity.

We are presented with many relativities in nature, such as the relativity of spatial orientation. It's important to bear in mind that this relativity does *not* assert that the equilibrium lengths of solid objects are unaffected by orientation; it merely asserts that all such lengths are affected by orientation in exactly the same proportion. It's conceivable that all solid objects are actually twice as long when oriented toward the Andromeda galaxy than when oriented perpendicular to that direction, but we have no way of knowing this.

Hence if we begin with the supposition that all objects are twice as long when pointed toward Andromeda, we could deduce that all lengths will appear to be independent of orientation, because they are all affected equally. But have we thereby "explained" the apparent isotropy of spatial lengths? Not at all, because the thing to be explained is the symmetry, i.e., why the lengths of all solid configurations, whether consisting of gold or wood, maintain exactly the same proportions, independent of their spatial orientations.

The Andromeda axis theory does not explain this physical symmetry. Instead, it explains something different, namely, why the Andromeda axis theory appears to be false even though it is (by supposition) true.

This is certainly a useful (indeed, essential) explanation for anyone who accepts, *a priori*, the truth of the Andromeda axis theory, but otherwise it is of very limited value.

Likewise if we accept absolute Galilean space and time as true concepts, *a priori*, then it is useful to understand why nature may appear to be Minkowskian, even though it is really (by supposition) Galilean. But what is the basis for the belief in the Galilean concept of space and time, as distinct from the Minkowskian concept, especially considering that the world appears to be Minkowskian?

Most physicists have concluded that there is no good answer to this question, and that it's preferable to study the world as it appears to be, rather than trying to rationalize "ancient habits". This does not imply a lack of interest in a deeper explanation for the effective symmetries of nature, but it does suggest that such explanations are most likely to come from studying those effective symmetries themselves, rather than from rationalizing why certain pre-conceived universal asymmetries would be undetectable.

A MORE PRACTICAL ARRANGEMENT

It's often overlooked that Einstein began his 1905 paper "On the Electrodynamics of Moving Bodies" by describing a system of coordinates based on a single absolute measure of time. He pointed out that we *could* assign time coordinates to each event

By using an observer located at the origin of the coordinate system, equipped with a clock, who coordinates the arrival of the light signal originating from the event to be timed and traveling to his position through empty space.

This is equivalent to Lorentz's conception of "true" time, provided the origin of the coordinate system is at "true" rest. However, for every frame of reference except the one at rest with the origin, these coordinates would not constitute an inertial coordinate system, because inertia would not be isotropic in terms of these coordinates, so Newton's laws of motion would not even be quasi-statically valid.

Furthermore, the selection of the origin is operationally arbitrary, and, even if the origin were agreed upon, there would be significant logistical difficulties in actually carrying out a coordination based on such a network of signals. Einstein says "We arrive at a much more *practical* arrangement by means of the following considerations".

In his original presentation of special relativity Einstein proposed two basic principles, derived from experience. The first is nothing other than Galileo's classical principle of inertial relativity, which asserts that for any material object in any state of motion there exists a system of space and time coordinates, called inertial coordinates, with respect to which the object is instantaneously at rest and inertia is homogeneous and isotropic (the latter being necessary for Newton's laws of motion to hold at least quasi-statically).

However, this principle alone is not sufficient to give a useful basis for evaluating physical phenomena. We must also have knowledge of how the description of events with respect to one system of inertial coordinates is related to the description of those same events with respect to another, relatively moving, system of coordinates. Rather than simply assuming a relationship based on some prior metaphysical conception of space and time, Einstein realized that the correct relationship between relatively moving systems of inertial coordinates could only be determined empirically.

He noted "the unsuccessful attempts to discover any motion of the earth relatively to the 'light medium", and since we define motion in terms of inertial coordinates, these experiments imply that the propagation of light is isotropic in terms of the very same class of coordinate systems for which mechanical inertia is isotropic.

On the other hand, all the experimental results that are consolidated into Maxwell's equations imply that the propagation speed of light (with respect to any inertial coordinate system) is independent of the state of motion of the emitting source. Einstein's achievement was to explain clearly how these seemingly contradictory facts of experience may be reconciled.

As an aside, notice that isotropy with respect to inertial coordinates is what we would expect if light was a stream of inertial corpuscles, whereas the independence of the speed of light from the motion of its source is what we would expect if light was a wave phenomenon.

This is the same dichotomy that we encounter in quantum mechanics, and it's not coincidental that Einstein wrote his seminal paper on light quanta almost simultaneously with his paper on the electrodynamics of moving bodies. He might actually have chosen to combine the two into a single paper discussing general heuristic considerations arising from the observed properties of light, and the reconciliation of the apparent dichotomy in the nature of light as it is usually understood.

From the empirical facts that light propagates isotropically with respect to every system of inertial coordinates (which is essentially just an extension of Galileo's principle of relativity), and that the speed of propagation of light with respect to any system of inertial coordinates is independent of the motion of the emitting source, it follows that the speed of light in invariant with respect to every system of inertial coordinates. From these facts we can deduce the correct relationship between relatively moving systems of inertial coordinates.

To establish the form of the relationships between this "more practical" class of coordinate systems (i.e., the class of inertial coordinate systems), Einstein notes that if x, y, z, t is a system of inertial coordinates, and a pulse of light is emitted from location x_0 along the x axis at time t_0 toward a distant location x_1, where it arrives and is reflected at time t_1, and if this reflected pulse is received back at location x_2 (the same as x_0) at time t_2 then $t_1 = (t_0 + t_2)/2$.

In other words, since light is isotropic with respect to the same class of coordinate systems in which mechanical inertia is isotropic, the light pulse takes the same amount of time, $(t_2 - t_1)/2$, to travel each way when expressed in terms of any system of inertial coordinates. By the same reasoning the spatial distance between the emission and reflection events is $x_1 - x_0 = c(t_2 - t_1)/2$.

Naturally the invariance of light speed with respect to inertial coordinates is implicit in the principles on which special relativity is based, but we must not make the mistake of thinking that this invariance is therefore tautological, or merely an arbitrary definition. Inertial coordinates are not arbitrary, and they are definable without explicit reference to the phenomenon of light. The real content of Einstein's principles is that light is an *inertial* phenomenon (despite its wavelike attributes).

The stationary ether of posited by Lorentz did not interact mechanically with ordinary matter at all, and yet we know that light conveys momentum to material objects. The coupling between the supposed ether and ordinary matter was always problematic for ether theories, and indeed for any classical wavelike theory of light. Einstein's paper on the photo-electric effect was a crucial step in recognizing the localized ballistic aspects of electromagnetic radiation, and this theme persists, just under the surface, of his paper on electrodynamics.

Oddly enough, the clearest statement of this insight came only as an afterthought, appearing in Einstein's second paper on relativity in 1905, in which he explicitly concluded that "radiation carries inertia between emitting and absorbing bodies". The point is that light conveys not only momentum, but inertia. For example, after a body has absorbed an elementary pulse of light, it has not only received a "kick" from the momentum of the light, but the internal inertia (i.e., the inertial mass) of the body has actually increased. Once it is posited that light is inertial, Galileo's principle of relativity automatically implies that light propagates isotropically from the source, regardless of the source's state of uniform motion.

Consequently, if we elect to use space and time coordinates in terms of which light speed is *not* isotropic (which we are certainly free to do), we will necessarily find that *no* inertial processes are isotropic. For example, we will find that two identical marbles expelled from a tube in opposite directions by an explosive charge located between them will not fly away at equal speeds, i.e., momentum will not be conserved.

Conversely, if we use ordinary mechanical inertial processes together with the conservation of momentum (and if we decline to assign any momentum or reaction to unobservable and/or immovable entities), we will necessarily arrive at clock synchronizations that are identical with those given by Einstein's light rays. Thus, Einstein's "more practical arrangement" is based on (and ensures) isotropy not just for light propagation, but for *all* inertial phenomena.

If a uniformly moving observer uses pairs of identical material objects thrown with equal force in opposite directions to establish spaces of simultaneity, he will find that his synchronization agrees with that produced by Einstein's assumed isotropic light rays.

The special attribute of light in this regard is due to the fact that, although light is inertial, it has no mass of its own, and therefore no rest frame. It can be regarded entirely as nothing but an interaction along a null interval between two massive bodies, the emitter and absorber. From this follows the indefinite metric of spacetime, and light's seemingly paradoxical combination of wavelike and inertial properties.

It's also worth noting that when Einstein invoked the operational definitions of time and distance based on light propagation, he commented that "we assume this definition of synchronization is free from contradictions, and possible for any number of points".

This is crucial for understanding why a set of definitions based on the propagation of light is tenable, in contrast with a similar set of definitions based on non-inertial signals, such as acoustical waves or postal messages. A set of definitions based on any non-inertial signal can't possibly preserve inertial isotropy. Of course, a signal requiring an ordinary material medium for its propagation would obviously not be suitable for a universal definition of time, because it would be inapplicable across regions devoid of that substance.

Moreover, even if we posited an omni-present substance, a signal consisting of (or carried by) any material substance would be unsuitable because such objects do not exhibit any particular fixed characteristic of motion, as shown by the fact that they can be brought to rest with respect to some inertial system of reference. Furthermore, if there exist any signals faster than those on which we base our definitions of temporal synchronization, those definitions will be easily falsified.

The fact that Einstein's principles are empirically viable at all, far from being vacuous or tautological, is actually somewhat miraculous.

In fact, if we were to describe the kind of physical phenomenon that would be required in order for us to have a consistent capability of defining a coherent basis of temporal synchronization for spatially separate events, clearly it could be neither a material object, nor a disturbance in a material medium, and yet it must exhibit some fixed characteristic quality of motion that exceeds the motion of any other object or signal.

We hardly have any right to expect, *a priori*, that such phenomenon exists. On the other hand, it could be argued that Einstein's second principle is just as classical as his first, because sight has always been the de facto arbiter of simultaneity.

Even in Galileo's day it was widely presumed that vision was instantaneous, so it automatically was taken to define simultaneity. The difference between this and the modern view is not so much the treatment of light as the means of defining simultaneity, but simply the realization that light propagates at a finite speed, and therefore the spacetime manifold is only partially ordered. The derivation of the Lorentz transformation presented in Einstein's

1905 paper is formally based on two empirically-based propositions, which he expressed as follows:

- The laws by which the conditions of physical systems change are independent of which of two coordinate systems in homogeneous translational movement relative to each other these changes in status are referred.
- Each ray of light moves in "the resting" coordinate system with the definite speed c, independently of whether this ray of light is emitted from a resting or moving body. Here speed = (optical path) / (length of time),

In the first of these propositions we are to understand that the "coordinate systems" are all such that Newton's laws of motion hold good (in a suitable limiting sense), This is crucial, because without this stipulation, the proposition is false. For example, coordinate systems related by Galilean transformations are "in homogeneous translational movement relative to each other", and yet the laws by which physical systems change (e.g., Maxwell's equations) are manifestly not independent of the choice of such coordinate systems.

So the restriction to coordinate systems in terms of which the laws of mechanics hold good is crucial. However, once we have imposed this restriction, the proposition becomes tautological, at least for the laws of mechanics.

The real content of Einstein's first "principle" is therefore the assertion that the other laws of physics (e.g., the laws of electrodynamics) hold good in precisely the same set of coordinate systems in terms of which the laws of mechanics hold good. Thus Einstein's first principle simply re-asserts Galileo's claim that *all* effects of uniform rectilinear motion can be "transformed away" by a suitable choice coordinate systems.

It might seems that Einstein's second principle is implied by the first, at least if Maxwell's equations are regarded as laws governing the changes of physical systems, because Maxwell's equations prescribe the speed of light propagation independent of the source's motion.

However, it's not clear *a priori* whether Maxwell's equations are valid in terms of relatively moving systems of coordinates, nor whether the permissivity of the vacuum is independent of the frame of reference in terms of which it is evaluated. Moreover, In 1905 Einstein already doubted the absolute validity of Maxwell's equations, having recently completed his paper on the photo-electric effect which introduced the idea of photons, i.e., light propagating as discrete packets of energy, a concept which cannot be represented as a solution of Maxwell's linear equations.

Einstein also realized that a purely electromagnetic theory of matter based on Maxwell's equations was impossible, because those equations by themselves could never explain the equilibrium of electric charge that constitutes a charged particle. "Only different, nonlinear field equations could possibly accomplish such a thing." This observation shows how unjustified was the "molecular force hypothesis" of Lorentz, according to which all the forces of nature were assumed to transform exactly as do electromagnetic forces as described by Maxwell's linear equations.

Knowing that the molecular forces responsible for the equilibrium of charged particles must necessarily be of a fundamentally different character than the forces of electromagnetism, and certainly knowing that the stability of matter may not even have a description in the form of a continuous field theory at all, it's clear that Lorentz's hypothesis has no constructive basis, and is simply tantamount to the adoption of Einstein's two principles.

Thus, Einstein's contribution was to recognize that "the bearing of the Lorentz transformation transcended its connection with Maxwell's equations and was concerned with the nature of space and time in general".

Instead of basing special relativity on an assumption of the absolutely validity of Maxwell's equations, Einstein based it on the particular characteristic exhibited by those equations, namely Lorentz invariance, that he intuited was the more fundamental principle, one that could serve as an organizing principle analogous to the conservation of energy in thermodynamics, and one that could encompass all physical laws, even if they turned out to be completely dissimilar to Maxwell's equations. Remarkably, this has turned out to be the case. Lorentz invariance is a key aspect of the modern theory of quantum electrodynamics, which replaced

Maxwell's equations. Of course, just as Einstein's first principle relies on the restriction to coordinate systems in which the laws of mechanics hold good, his second principle relies crucially on the requirement that time intervals are "to be understood in the sense of the definition". And, again, once this condition is recognized, the principle itself becomes tautological, although in this case the tautology is complete.

The second principle states that light always propagates at the speed c, assuming we define the time intervals in accord, which defines time intervals as whatever they must be in order for the speed of light to be c.

This unfortunately has led some critics to assert that special relativity is purely tautological, merely a different choice of conventions. Einstein's presentation somewhat obscures the real physical content of the theory, which is that mechanical inertia and the propagation speed of light are isotropic and invariant with respect to precisely the same set of coordinate systems. This is a non-trivial fact. It then remains to determine how these distinguished coordinate systems are related to each other.

Although Einstein explicitly highlighted just two principles as the basis of special relativity in his 1905 paper (consciously patterned after the two principle of thermodynamics), his derivation of the Lorentz transformation also invoked "the properties of *homogeneity* that we attribute to space and time" to establish the linearity of the transformations. In addition, he tacitly assumed *spatial isotropy*, i.e., that there is no preferred direction in space, so the intrinsic properties of ideal rods and clocks do not depend on their spatial orientations.

Lastly, he assumed *memorylessness*, i.e., that the extrinsic properties of rods and clocks may be functions of their current positions and states of motion, but not of their previous positions or states of motion. This last assumption is needed to exclude the possibility that every elementary particle may somehow "remember" its entire history of accelerations, and thereby "know" its present absolute velocity relative to a common fixed reference. (Einstein explicitly listed these extra assumptions in an exposition written in 1920. He may have gained an appreciation of the importance of the independence of measuring rods and clocks from their past history after considering Weyl's unified field theory, which Einstein rejected precisely because it violated this premise).

The actual detailed derivation of the Lorentz transformation presented in Einstein's 1905 paper is somewhat obscure and circuitous, but it's worthwhile to follow his reasoning, partly for historical interest, and partly to contrast it with the more direct and compelling derivations that will be presented in subsequent sections.

Following Einstein's original derivation, we begin with an inertial (and Cartesian) coordinate system called K, with the coordinates x, y, z, t, and we posit another system of inertial coordinates denoted as k, with the coordinates ξ, η, ζ, τ. The spatial axes of these two systems are aligned, and the spatial origin of k is moving in the positive x direction with speed v in terms of K.

We then consider a particle at rest in the k system, and note that for such a particle the x and t coordinates (i.e., the coordinates in terms of the K system) are related by x' = x – vt for some constant x'. We also know the y and z coordinates of such a particle are constant. Hence each stationary spatial position in the k system corresponds to a set of three constants (x', y, z), and we can also assign the time coordinate t to each event.

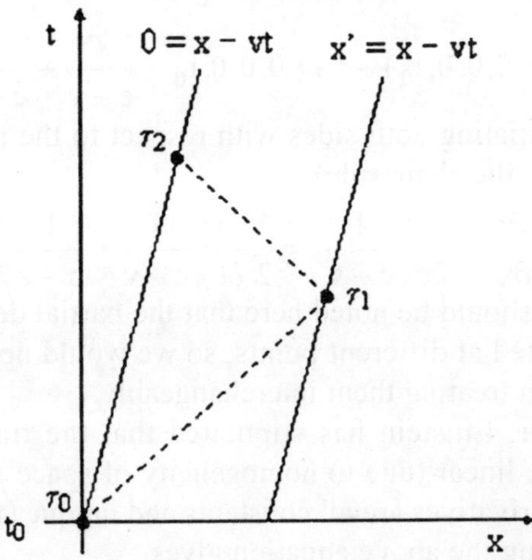

Interestingly, the system of variables x', y, z, t constitute a complete coordinate system, related to the original system K by a Galilean transformation x' = x-vt, y'=y, z'=z, t'=t. Thus, just as

Lorentz did in 1892, Einstein began by essentially applying a Galilean transformation to the original "rest frame" coordinates to give an intermediate system of coordinates, although Einstein's paper makes it clear that this is not an inertial coordinate system.

Now we consider the values of the τ coordinate of the k system as a function of x', y, z, t for any stationary point in the k system. Suppose a pulse of light is emitted from the origin of the k system in the positive x direction at time τ_0, it reaches the point corresponding to x', y, z at time τ_1, where it is reflected, arriving back at the origin of the k system at time τ_2. This is depicted in the figure back.

Recall that the ξ, η, ζ, τ coordinates are defined as inertial coordinates, meaning that inertia is homogeneous and isotropic in terms of these coordinates. Also, all experimental evidence (such as all "the unsuccessful attempts to discover any motion of the earth relatively to the 'light medium'") indicates that the speed of light is isotropic in terms of any inertial coordinate system. Therefore, we have $\tau_1 = (\tau_0 + \tau_2)/2$, so the τ coordinate as a function of x', y, z, t satisfies the relation

$$\tau\left(x', 0, 0, t_0 + \frac{x'}{c-v}\right)$$

$$= \frac{1}{2}\left[\tau(0,0,0,t_0) + \tau + \left(0,0,0,t_0 + \frac{x'}{c+v} + \frac{x'}{c+v}\right)\right]$$

Differentiating both sides with respect to the parameter x', we get (using the chain rule)

$$\frac{\partial \tau}{\partial x'} + \frac{\partial \tau}{\partial t}\left(\frac{1}{c-v}\right) = \frac{1}{2}\frac{\partial \tau}{\partial t}\left(\frac{1}{c-v} + \frac{1}{c-v}\right)$$

Now, it should be noted here that the partial derivatives are being evaluated at different points, so we would not, in general, be justified in treating them interchangeably.

However, Einstein has stipulated that the transformation equations are linear (due to homogeneity of space and time), so the partial derivatives are all constants and unique (for any given v). Simplifying the above equation gives

$$\frac{\partial \tau}{\partial x'} = \left(\frac{v}{c^2 - v^2}\right)\frac{\partial \tau}{\partial t}$$

At this point, Einstein alludes to analogous reasoning for the y and z directions, but doesn't give the details. Presumably we are to consider a pulse of light emanating from the origin and reflecting at a point x' = 0, y, z = 0, and returning to the origin. In this case the isotropy of light propagation in terms of inertial coordinates implies

$$\tau\left(0, y, 0, t_0 + \frac{y}{\sqrt{c^2 - v^2}}\right)$$
$$= \frac{1}{2}\left[\tau(0,0,0,t_0) + \tau\left(0,0,0,t_0 + \frac{2y}{\sqrt{c^2 - v^2}}\right)\right]$$

In this equation we have made use of the fact that the y component of the speed of the light pulse (in terms of the K system) as it travels in either direction between these points, which are stationary in the k system, is $(c^2 - v^2)^{1/2}$. Differentiating both sides with respect to y, we get

$$\frac{\partial\tau}{\partial y} + \frac{\partial\tau}{\partial t}\frac{2}{\sqrt{c^2 - v^2}} = \frac{1}{2}\frac{\partial\tau}{\partial t}\frac{2}{\sqrt{c^2 - v^2}}$$

and therefore “τ/”y = 0. The same reasoning shows that “τ/”z = 0. Now the total differential of τ(x', y, z, t) is, by definition

$$d\tau = \frac{\partial\tau}{\partial x'}dx' + \frac{\partial\tau}{\partial y}dy + \frac{\partial\tau}{\partial z}dz + \frac{\partial\tau}{\partial t}dt$$

and we know the partial derivatives with respect to y and z are zero, and the partial derivatives with respect to x' and t are in a known ratio, so for any given v we can write

$$d\tau = a(v)\left(dt - \frac{v}{c^2 - v^2}dx'\right)$$

where a(v) is as yet an undetermined function. Incidentally, Einstein didn't write this expression in terms of differentials, but he did state that he was “letting x' be infinitesimally small”, so he was essentially dealing with differentials. On the other hand, the distinction between differentials and finite quantities matters little in this context, because the relations are linear, and hence the partial derivatives are constants, so the differentials can be trivially integrated. Thus we have

$$\tau = a(v)\left(t - \frac{v}{c^2 - v^2}x'\right)$$

Einstein then used this result to determine the transformation equations for the spatial coordinates. The ξ coordinate of a pulse of light emitted from the origin in the positive x direction is related to the τ coordinate by $\xi = c\tau$ (since experience has shown that light propagates with the speed c in all directions when expressed in terms of any system of inertial coordinates). Substituting for τ from the preceding formula gives, for the ξ coordinate of this light pulse, the expression

$$\xi = ca(v)\left(t - \frac{v}{c^2 - v^2}x'\right)$$

We also know that, for this light pulse, the parameters t and x' are related by $t = x'/(c-v)$, so we can substitute for t in the above expression and simplify to give the relation between ξ and x' (both of which, we remember, are constants for any point at rest in k)

$$\xi = a(v)\left(\frac{c^2}{c^2 - v^2}\right)x'$$

We can choose x' to be anything we like, so this represents the general relation between these two parameters. Similarly the η coordinate of a pulse of light emanating from the origin in the η direction is

$$\eta = c\tau = ca(v)\left(t - \frac{v}{c^2 - v^2}x'\right)$$

but in this case we have x' = 0 and, as noted previously, $t = y/(c^2 - v^2)^{1/2}$, so we have

$$\eta = a(v)\left(\frac{c}{\sqrt{c^2 - v^2}}\right)y$$

and by the same token

$$\zeta = a(v)\left(\frac{c}{\sqrt{c^2 - v^2}}\right)z$$

If we define the function

$$\phi(v) = a(v)\left(\frac{c}{\sqrt{c^2 - v^2}}\right)$$

and substitute x – vt for x', the preceding results can be summarized as

$$\tau = \phi(v)\frac{t - vx/c^2}{\sqrt{1-(v/c)^2}} \quad \xi = \phi(v)\frac{x - vt}{\sqrt{1-(v/c)^2}} \quad \eta = \phi(v)y \quad \zeta = \phi(v)z$$

At this point Einstein observes that a sphere of light expanding with the speed c in terms of the unprimed coordinates transforms to a sphere of light expanding with speed c in terms of the double-primed coordinates. In other words,

$$x^2 + y^2 + z^2 = c^2t^2 \quad \text{implies} \quad \xi^2 + \eta^2 + \zeta^2 = c^2\tau^2$$

As Einstein says, this "shows that our two fundamental principles are compatible", i.e., it is possible for light to propagate isotropically with respect to two relatively moving systems of inertial coordinates, provided we allow the possibility that the transformation from one inertial coordinate system to another is not exactly as Galileo and Newton surmised.

To complete the derivation of the Lorentz transformation, it remains to determine the function $\phi(v)$. To do this, Einstein considers a two-fold application of the transformation, once with the speed v in the positive x direction, and then again with the speed v in the negative x direction.

The result should be the identity transformation, i.e., we should get back to the original coordinate system. (Strictly speaking, this assumes the property of "memorylessness"). It's easy to show that if we apply the above transformation twice, once with parameter v and once with parameter –v, each coordinate is $\phi(v)\phi(-v)$ times the original coordinate, so we must have

$$\phi(v)\phi(-v) = 1$$

Finally, Einstein concludes by "inquiring into the signification of $\phi(v)$". He notes that a segment of the η axis moving with speed v perpendicular to its length (i.e., in the positive x direction) has the length $y = \eta/\phi(v)$ in terms of the K system coordinates, and by "reasons of symmetry" (i.e., spatial isotropy) this must equal

$\eta/\phi(-v)$, because it doesn't matter whether this segment of the y axis is moving in the positive or the negative x direction. Consequently we have $\phi(v) = \phi(-v)$, and therefore $\phi(v) = 1$, so he arrives at the Lorentz transformation

$$\tau = \frac{t - vx/c^2}{\sqrt{1-)(v/c)^2}} \quad \xi = \frac{x - vt}{\sqrt{1-(v/c)^2}} \quad \eta = y \quad \zeta = z$$

This somewhat laborious and awkward derivation is interesting in several respects. For one thing, one gets the impression that Einstein must have been experimenting with various methods of presentation, and changed his nomenclature during the drafting of the paper.

For example, at one point he says "a is a function $\phi(v)$ at present unknown", but subsequently a(v) and $\phi(v)$ are defined as different functions.

At another point he defines x' as a Galilean transform of x (without explicitly identifying it as such), but subsequently uses the symbol x' as part of the inertial coordinate system resulting from the two-fold application of the Lorentz transformation.

In addition, he somewhat tacitly makes use of the invariance of the light-like relation $x^2 + y^2 = c^2t^2$ in his derivation of the transformation equations for the y coordinate, but doesn't seem to realise that he could just as well have invoked the invariance of $x^2 + y^2 + z^2 = c^2t^2$ to make short work of the entire derivation. Instead, he presents this invariance as a *consequence* of the transformation equations – despite the fact that he has tacitly used the invariance as the *basis* of the derivation (which of course he was entitled to do, since that invariance simply expresses his "light principle"). Perhaps not surprisingly, some readers have been confused as to the significance of the functions a(v) and $\phi(v)$. For example, in a review of Einstein's paper, A. I. Miller writes

Then, without prior warning Einstein replaced a(v) with $\phi(v)/(1-(v/c)^2)^{1/2}$...But why did Einstein make this replacement? It seems as if he knew beforehand the correct form of the set of relativistic transformations... How did Einstein know that he had to make [this substitution] in order to arrive at those space and time transformations in agreement with the postulates of relativity?

This suggests a misunderstanding, because the substitution

in question is purely formal, and has no effect on the content of the equations. The transformations that Einstein had derived by that point, prior to replacing a(v), were already consistent with the postulates of relativity.

It is simply more convenient to express the equations in terms of $\phi(v)$, which is the entire coefficient of the transformations for y and z. One naturally expects this coefficient to equal unity.

Even aside from the inadvertent changes in nomenclature, Einstein's derivation is undeniably clumsy, especially in first applying what amounts to a Galilean transformation, and then deriving the further transformation needed to arrive at a system of inertial coordinates. It's clear that he was influenced by Lorentz's writings, even to the point of using the same symbol β for the quantity $1/(1-(v/c)^2)^{1/2}$, which Lorentz used in his 1904 paper. (Oddly enough, many years later Einstein wrote to Carl Seelig that in 1905 he had known only of Lorentz's 1895 paper, but not his subsequent papers, and none of Poincare's papers on the subject).

In a review article published in 1907 Einstein had already adopted a more economical derivation, dispensing with the intermediate Galilean system of coordinates, and making direct use of the lightlike invariant expression, similar to the standard derivation presented in most introductory texts today.

To review this now standard derivation, consider (again) Einstein's two systems of inertial coordinates K and k, with coordinates denoted by (x, y, z, t) and (ξ, η, ζ, τ) respectively, and oriented so that the x and ξ axes coincide, and the xy plane coincides with the $\xi\eta$ plane. Also, as before, the system k is moving in the positive x direction with fixed speed v relative to the system K, and the origins of the two systems momentarily coincide at time $t = \tau = 0$.

According to the principle of homogeneity, the relationship between the two sets of coordinates must be linear, so there must be constants A_1 and A_2 (for a given v) such that $\xi = A_1 x + A_2 t$. Furthermore, if an object is stationary relative to k, and if it passes through the point (x, t) = (0, 0), then it's position in general satisfies x = vt, from the definition of velocity, and the ξ coordinate of that point with respect to the k system is 0. Therefore we have ξ

$= A_1(vt) + A_2\, t = 0$. Since this must be true for non-zero t, we must have $A_1\, v + A_2 = 0$, and so $A_2 = -A_1\, v$. Consequently, there is a single constant A (for any given v) such that $\xi = A(x-vt)$. Similarly there must be constants B and C such that $\eta = By$ and $\zeta = Cz$. Also, invoking isotropy and homogeneity, we know that τ is independent of y and z, so it must be of the form $\tau = Dx + Et$ for some constants D and E (for a given v). It only remains to determine the values of the constants A, B, C, D, and E in these expressions. Suppose at the instant when the spatial origins of K and k coincide a spherical wave of light is emitted from their common origin. At a subsequent time t in the first frame of reference the sphere of light must be the locus of points satisfying the equation

$$x^2 + y^2 + z^2 = c^2 t^2$$

and likewise, according to our principles, in the second frame of reference the spherical wave at time τ must be the locus of points described by

$$\xi^2 + \eta^2 + \zeta^2 = c^2 \tau^2$$

Substituting from the previous expressions for the k coordinates into this equation, we get

$$(A(x - vt))^2 + (By)^2 + (Cz)^2 = c^2 (Dx + Et)^2$$

Expanding these terms and rearranging gives

$$\left(A^2 - c^2 D^2\right) x^2 + \left(B^2\right) y^2 + \left(C^2\right) z^2 - 2\left(A^2 v + DEc^2\right) xt$$

$$= c^2 \left[E^2 - A^2 (v/c)^2\right] t^2$$

The assumption that light propagates at the same speed in both frames of reference implies that a simultaneous spherical shell of light in one frame is also a simultaneous spherical shell of light in the other frame, so the coefficients of equation must be proportional to the coefficients of equation.

Strictly speaking, the constant of proportionality is arbitrary, representing a simple re-scaling, so we are free to impose an additional condition, namely, that the transformation with parameter +v followed by the transformation with parameter –v yields the original coordinates, and by the isotropy of space these two transformations, which differ only in direction, must have the same constant of proportionality. Thus the corresponding coefficients of equations must not only be proportional, they must be equal, so we have

$$A^2c^2D^2 = 1 \quad B^2 = 1 \quad C^2 = 1$$

$$2(A^2v + DEc^2) = 0 \quad E^2 - A^2(v/c)^2 = 1$$

Clearly we can take B = C = 1 (rather than −1, since we choose not to reflect the y and z directions). Dividing the 4th of these equations by 2, we're left with the three equations in the three unknowns A, D, and E:

$$A^2 - c^2D^2 = 1 \quad A^2v + DEc^2 = 0 \quad E^2 - A^2(v/c)^2 = 1$$

Solving the first equation for A^2 and substituting this into the 2nd and 3rd equations gives

$$(1 + c^2D^2)v + DEc^2 = 0 \quad E^2 - (1 + c^2D^2)(v/c)^2 = 1$$

Solving the first for E and substituting into the 2nd gives a single quadratic equation in D, with the roots

$$D = \frac{\pm v/c^2}{\sqrt{1-(v/c)^2}}$$

Substituting this into either of the previous equations and solving the resulting quadratic for E gives

$$E = \frac{\mp 1}{\sqrt{1-(v/c)^2}}$$

Note that the equations require opposite signs for D and E. Now, for small values of v/c we expect to find E approaching +1 (as in Galilean relativity), so we choose the positive root for E and the negative root for D. Finally, from the relation $A^2 - c^2 D^2 = 1$ we get

$$A = \frac{\pm 1}{\sqrt{1-(v/c)^2}}$$

and again we select the positive root. Consequently we have the Lorentz transformation

$$\xi = A(x - vt) = \frac{x - vt}{\sqrt{1-(v/c)^2}}$$

$$\eta = By = y \qquad \zeta = Cz = z$$

$$\tau = Dx + Et\frac{-xv/c^2 + t}{\sqrt{1-(v/c)^2}}$$

Naturally with this transformation we can easily verify that

$$x^2 + y^2 + z^2 - c^2t^2 = \xi^2 + \eta^2 + \zeta^2 - c^2\tau^2$$

so this quantity is the squared "absolute distance" from the origin to the point with K coordinates (x, y, z, t) and the corresponding k coordinates (ξ, η, ζ, τ), which confirms that the absolute spacetime interval between two points is the same in both frames.

Notice that equations already implied this relation for null intervals. In other words, the original premise was that if $x^2 + y^2 + z^2 - c^2t^2$ equals zero, then $\xi^2 + \eta^2 + \zeta^2 - c^2\tau^2$ also equals zero. The above reasoning show that a consequence of this premise is that, for *any* arbitrary real number s^2, if $x^2 + y^2 + z^2 - c^2t^2$ equals s^2, then $\xi^2 + \eta^2 + \zeta^2 - c^2\tau^2$ also equals s^2. Therefore, this quadratic form represents an absolute invariant quantity associated with the interval from the origin to the event (x, y, z, t).

STAIRCASE WIT

In retrospect, it's easy to see that the Galilean notion of space and time was not free of conceptual difficulties. In 1908 Minkowski delivered a famous lecture in which he argued that the relativistic phenomena described by Lorentz and clarified by Einstein might have been inferred from first principles long before, if only more careful thought had been given to the foundations of classical geometry and mechanics.

He pointed out that special relativity arises naturally from the reconciliation of two physical symmetries that we individually take for granted. One is spatial isotropy, which asserts the equivalence of all physical phenomena under linear transformations such as $x' = ax - by$, $y' = bx + ay$, $z' = z$, $t' = t$, where $a^2 + b^2 = 1$. It's easy to verify that transformations of this type leave all quantities of the form $x^2 + y^2 + z^2$ invariant. The other is Galilean relativity, which asserts the equivalence of all physical phenomena under transformations such as x' = x – vt, y' = y, z' = z, t' = t, where v is a constant. However, these transformations obviously do not leave the quantity $x^2 + y^2 + z^2$ invariant, because they involve the time coordinate as well as the space coordinates.

In addition, we notice that the rotational transformations maintain the orthogonality of the coordinate axes, whereas the lack of an invariant measure for the Galilean transformations prevents us from even assigning a definite meaning to "orthogonality" between the time and space coordinates. Since the velocity transformations leave the laws of physics unchanged, Minkowski reasoned, they ought to correspond to some invariant physical quantity, and their determinants ought to be unity.

Clearly the invariant must involve the time coordinate, and hence the units of space and time must be in some fixed non-singular relation to each other, with a conversion factor that we can normalize to unity. Also, since we cannot go backwards in time, the space axis must not be rotated in the same direction as the time axis by a velocity transformation, so the velocity transformations ought to be of the form x' = ax – bt, y'=y, z'=z, t' = bx – at, where $a^2 - b^2 = 1$. Combining this with the requirements b/a = v, we arrive at the transformation

$$x' = \frac{x - vt}{\sqrt{1 - v^2}} \quad y' = y \quad z' = z \quad t' = \frac{t - vx}{\sqrt{1 - v^2}}$$

which leaves invariant the quantity $x^2 + y^2 + z^2 - t^2$. The rotational transformations also leave this same quantity invariant, so this appears to be the most natural (and almost the only) way of reconciling the observed symmetries of physical phenomena. Hence from simple requirements of rational consistency we could have arrived at the Lorentz transformation. As Minkowski said

> Such a premonition would have been an extraordinary triumph for pure mathematics. Well, mathematics, though it now can display only staircase wit, has the satisfaction of being wise after the event... to grasp the far-reaching consequences of such a metamorphosis of our concept of nature.

physical entity had constant values (for a given position and orientation), but of course this assumption is valid only if the entities are stationary. If an object is in motion (relative to the system of coordinates), then the coordinates of its endpoints are variable functions of time, so instead of the constant x_1 we have a function $x_1(t)$, and likewise for the other coordinates.

It's natural to ask whether the symmetry of equation is still applicable to objects in motion. Clearly if we allow the individual coordinate functions to be evaluated at unequal times then the symmetry does not apply. However, if all the coordinate functions are evaluated for the same time, experience teaches us that equation does apply to objects in motion.

This is the second of our two commonplace symmetries, the apparent fact that the sum of the squares of the orthogonal components of the spatial interval between the two ends of a solid entity is invariant for all states of uniform motion, with the understanding that the coordinates are all evaluated at the same time.

To express this symmetry more precisely, let x_1, y_1, z_1 denote the spatial coordinates of one end of a solid physical entity at time t_1, and let x_2, y_2, z_2 denote the spatial coordinates of the other end at time t_2. Then the quantity expressed by equation is invariant for any position, orientation, and state of uniform motion *provided* $t_1 = t_2$. However, just as the spatial part of the symmetry is not valid for arbitrary spatial coordinate systems, the temporal part is not valid for arbitrary time coordinates. Recall that the spatial isotropy of the quantity expressed by equation is valid only if the space coordinates x, y, z are mutually orthogonal. Likewise, the combined symmetry covering states of uniform motion is valid only if the time component t is mutually orthogonal to each of the space coordinates.

The question then arises as to how we determine whether coordinate axes are mutually orthogonal. We didn't pause to consider this question when we were dealing only with the three spatial coordinates, but even for the three space axes the question is not as trivial as it might seem.

The answer relies on the concept of "distance" defined by the quantity s in equation. According to Euclid, two lines intersecting at the point P are perpendicular if and only if each point of one line is equidistant from the two points on the other line that are equidistant from P. Unfortunately, this reasoning involves a circular argument, because in order to determine whether two lines are orthogonal, we must evaluate distances between points on those lines using an equation that is valid only if our coordinate axes are orthogonal.

By this reasoning, we could conjecture that any two obliquely intersecting lines are orthogonal, and then use equations with coordinates based on those lines to confirm that they are indeed orthogonal according to Euclid's definition. But of course the physical objects of our experience would not exhibit spatial isotropy in terms of these coordinates. This illustrates that we can only establish the physical orthogonality of coordinate axes based on physical phenomena. In other words, we construct orthogonal coordinate axes operationally, based on the properties of physical entities.

For example, we define an orthogonal system of coordinates in such a way that a certain spatially extended physical entity is isotropic. Then, by definition, this physical entity is isotropic with respect to these coordinates, so again the reasoning is circular. However, the physical significance of these coordinates and the associated spatial isotropy lies in the empirical fact that all other physical entities (in the class of "solids") exhibit spatial isotropy in terms of this same system of coordinates.

Next we need to determine a time axis that is orthogonal to each of the space axes. In common words, this amounts to synchronizing the times at spatially separate locations. Just as in the case of the spatial axes, we can establish physically meaningful orthogonality for the time axis only operationally, based on some reference physical phenomena.

As we've seen, orthogonality between two lines is determined by the distances between points on those lines, so in order to determine a time axis orthogonal to a space axis we need to evaluate "distances" between points that are separated in time as well as in space. Unfortunately, equation defines distances only between points at the same time. Evidently to establish orthogonality between space and time axes we need a physically meaningful measure of *space-time* distance, rather than merely spatial distance. Another physical symmetry that we observe in nature is the symmetry of temporal translation. This refers to the fact that for a certain class of physical processes the duration of the process is independent of the absolute starting time. In other words, letting t_1 and t_2 denote the times of the two ends of the process, the quantity

$$\tau^2 = (t_2 - t_1)^2$$

is invariant under translation of the starting time t_1. This is exactly analogous to the symmetry of a class of physical objects under spatial translations. However, we have seen that the spatial symmetries are valid only if the time coordinates t_1 and t_2 are the same, so we should recognize the possibility that the physical symmetry expressed by the invariance is valid only when the spatial coordinates of events 1 and 2 are the same.

Of course, this can only be determined empirically. Somewhat surprisingly, common experience suggests that the values of τ^2 for a certain class of physical processes actually are invariant even if the spatial positions of events 1 and 2 are different… at least to within the accuracy of common observation and for differences in positions that are not too great.

Likewise we find that, for just about any time axis we choose, such that some material object is at rest in terms of the coordinate system, the spatial symmetries indicated by equation apply, at least within the accuracy of common observation and for objects that are not moving too rapidly. This all implies that the ratio of spatial to temporal units of distance is extremely great, if not infinite.

If the ratio is infinite, then *every* time axis is orthogonal to every space axis, whereas if it is finite, any change of the direction of the time axis requires a corresponding change of the spatial axes in order for them to remain mutually perpendicular.

The same is true of the relation between the space axes themselves, i.e., if the scale factor between (say) the x and the y coordinates was infinite, then those axes would always be perpendicular, but since it is finite, any rotation of the x axis (about the z axis) requires a corresponding rotation of the y axis in order for them to remain orthogonal.

It is perhaps conceivable that the scale factor between space and time could be infinite, but it would be very incongruous, considering that the time axis can have spatial components. We have no means of quantifying the absolute separation between two non-simultaneous events. The spatial separation between *non-simultaneous* events separated by a time increment Δt is totally undefined, because there exist perfectly valid reference frames in which two non-simultaneous events are at precisely the same spatial location, and other frames in which they are arbitrarily far apart.

Still, in all of those frames (according to Galilean relativity), the time interval remains Δt. Thus, there is no definite *combined* spatial and temporal separation – despite the fact that we clearly intuit a definite physical difference between our distance from "the office tomorrow" and our distance from "the Andromeda galaxy tomorrow". Admittedly we could postulate a universal preferred reference frame for the purpose of assessing the complete separations between events, but such a postulate is entirely foreign to the logical structure of Galilean space and time, and has no operational significance.

So, we are led to suspect that there is a finite (though perhaps very large) scale factor c between the units of space and time, and that the physical symmetries we've been discussing are parts of a larger symmetry, comprehending the spatial symmetries expressed and the temporal symmetries.

On the other hand, we do not expect spacelike intervals and timelike intervals to be directly conformable, because we cannot turn around in time as we can in space.

The most natural supposition is that the squared spacelike intervals and the squared timelike intervals have opposite signs, so that they are mutually "imaginary" (in the numerical sense). Hence our proposed invariant quantity for a suitable class of repeatable physical processes extending uniformly from event 1 to event 2 is

$$s^2 = (x_2 - x_1)^2 + (y_2 - y_1)^2 + (z_2 - z_1)^2 - c^2(t_2 - t_1)^2$$

This is the conventional form for spacelike intervals, whereas the negative of this quantity, denoted by τ^2, is used to signify timelike intervals. This quantity is invariant under any combination of spatial rotations and changes in the state of uniform motion, as well as simple translations of the origin in space and/or time.

The algebraic group of all transformations (not counting reflections) that leave this quantity invariant is called the Poincare group, in recognition of the fact that it was first described in Poincare's famous "Palermo" paper, dated July 1905.

Equation is not positive-definite, which means that even though it is a squared quantity it may have a negative value, and of course it vanishes along the path of a light pulse. Noting that squared times and squared distances have opposite signs, Minkowski remarked that

Thus the essence of this postulate may be clothed mathematically in a very pregnant manner in the mystic formula

$$3.10^5 \ \text{km} = \sqrt{-1} \ \text{sec s}$$

On this basis equation can be re-written in a way that is formally symmetrical in the space and time coordinates, but of course the invariant quantity remains non-positive-definite. The significance of this "mystic formula" continues to be debated, but it does provide an interesting connection to quantum mechanics.

As an aside, note that measurements of physical objects in various orientations are not sufficient to determine the "true" lengths in any metaphysical absolute sense. If all physical objects were, say, twice as long when oriented in one particular absolute direction than in the perpendicular directions, and if this anisotropy affected all physical phenomena equally, we could never detect it, because our rulers would be affected as well.

Thus, when we refer to a physical symmetry (such as the isotropy of space), we are referring to the fact that all physical phenomena are affected by some variable (such as spatial orientation) in exactly the same way, not that the phenomena bear any particular relationship with some metaphysical standard. From this perspective we can see that the Lorentzian approach to "explaining" the (apparent) symmetries of space-time does nothing to actually explain those symmetries; it is simply a rationalization of the discrepancy between those empirical symmetries and an *a priori* metaphysical standard that does not possess those symmetries.

In any case, we've seen how a slight (for most purposes) modification of the relationship between inertial coordinate systems leads to the invariant quantity

$$(dx)^2 + (dx)^2 + (dx)^2 - c^2(dt)^2$$

For any fixed value of the constant c, we will denote by G_c the group of transformations that leave this quantity unchanged. If we let c go to infinity, the temporal increment dt must be invariant, leaving just the original Euclidean group for the spatial increments. Thus the space and time components are de-coupled, in accord with Galilean relativity. Minkowski called this limiting case G_∞, and remarked that

Since G_c is mathematically much more intelligible than G_∞, it looks as though the thought might have struck some mathematician, fancy-free, that after all, as a matter of fact, natural phenomena do not possess invariance with the group G_∞, but rather with the group G_c, with c being finite and determinate, but in ordinary units of measure extremely great.

Minkowski is here clearly suggesting that Lorentz invariance might have been deduced from *a priori* considerations, appealing to mathematical "intelligibility" as a criterion for the laws of nature. Einstein himself eschewed the temptation to retroactively deduce Lorentz invariance from first principles, choosing instead to base his original presentation of special relativity on two empirically-founded principles, the first being none other than the classical principle of relativity, and the second being the proposition that the speed of light is the same with respect to any system of inertial coordinates, independent of the motion of the source.

This second principle often strikes people as arbitrary and unwarranted, and there have been numerous attempts to deduce it from some more fundamental principle. For example, it's been argued that the light speed postulate is actually redundant to the relativity principle itself, since if we regard Maxwell's equations as fundamental laws of physics, *and* we regard the permeability μ_0 and permittivity ε_0 of the vacuum as invariant constants of those laws in any uniformly moving frame of reference, then it follows that the speed of light in a vacuum is $c = 1/\sqrt{\mu_0 \varepsilon_0}$ with respect to every uniformly moving system of coordinates. The problem with this line of reasoning is that Maxwell's equations are *not* valid when expressed in terms of an arbitrary uniformly moving system of coordinates.

In particular, they are not invariant under a Galilean transformation - despite the fact that systems of coordinates related by such a transformation are uniformly moving with respect to each other. (Maxwell himself recognized that the equations of electromagnetism, unlike Newton's equations of mechanics, were not invariant under Galilean "boosts"; in fact he proposed various experiments to exploit this lack of invariance in order to measure the "absolute velocity" of the Earth relative to the aluminiferous ether.

Furthermore, we cannot assume, *a priori*, that μ_0 and ε_0 are invariant with respect to changes in reference frame. Actually μ_0 is an assigned value, but ε_0 must be measured, and the usual means of empirically determining ε_0 involve observations of the force between charged plates. Maxwell clearly believed these measurements must be made with the apparatus "at rest" with respect to the ether in order to yield the true and isotropic value of ε_0.

In sections 768 and 769 of Maxwell's Treatise he discussed the ratio of electrostatic to electromagnetic units, and predicted that two parallel sheets of electric charge, both moving in their own planes in the same direction with velocity c (supposing this to be possible) would exert no net force on each other.

If Maxwell imagined himself moving along with these charged plates and observing no force between them, he obviously did not expect the laws of electrostatics to be applicable. (This is analogous to Einstein's famous thought experiment in which he imagined moving along side a relatively "stationary" pulse of light).

According to Maxwell's conception, if measurements of ε_0 are performed with an apparatus traveling at some significant fraction of the speed of light, the results would not only differ from the result at rest, they would also vary depending on the orientation of the plates relative to the direction of the absolute velocity of the apparatus.

Of course, the efforts of Maxwell and others to devise empirical methods for measuring the absolute rest frame (either by measuring anisotropies in the speed of light or by detecting variations in the electromagnetic properties of the vacuum) were doomed to failure, because even though it's true that the equations of electromagnetism are *not* invariant under Galilean transformations, it is also true that those equations *are* invariant with respect to every system of inertial coordinates.

Maxwell (along with everyone else before Einstein) would have regarded those two propositions as logically contradictory, because he assumed inertial coordinate systems are related by Galilean transformations. Einstein was the first to recognize that this is not so, i.e., that relatively moving inertial coordinate systems are actually related by Lorentz transformations.

Maxwell's equations are suggestive of the invariance of c only because of the added circumstance that we are unable to physically identify any particular frame of reference for the application of those equations. (Needless to say, the same is not true of, for example, the Navier-Stokes equation for a material fluid medium).

The most readily observed instance of this inability to single out a unique reference frame for Maxwell's equations is the empirical invariance of light speed with respect to every inertial system of coordinates, from which we can infer the invariance of ε_0. Hence attempts to deduce the invariance of light speed from Maxwell's equations are fundamentally misguided.

Furthermore, we know (as did Einstein) that Maxwell's equations are not fundamental, since they don't encompass quantum photo-electric effects (for example), whereas the Minkowski structure of spacetime (representing the invariance of the local characteristic speed of light) evidently *is* fundamental, even in the context of quantum electrodynamics.

This strongly supports Einstein's decision to base his kinematics on the light speed principle itself. Another argument that is sometimes advanced in support of the second postulate is based on the notion of causality. If the future is to be determined by (and only by) the past, then (the argument goes) no object or information can move infinitely fast, and from this restriction people have tried to infer the existence of a finite upper bound on speeds, which would then lead to the Lorentz transformations.

One problem with this line of reasoning is that it's based on a principle (causality) that is not unambiguously self-evident. Indeed, if certain objects could move infinitely fast, we might expect to find the universe populated with large sets of indistinguishable particles, all of which are really instances of a small number of prototypes moving infinitely fast from place to place, so that they each occupy numerous locations at all times.

This may sound implausible until we recall that the universe actually *is* populated by apparently indistinguishable electrons and protons, and in fact according to quantum mechanics the individual identities of those particles are ambiguous in many circumstances. John Wheeler once seriously toyed with the idea that there is only a single electron in the universe, weaving its way back and forth through time. Admittedly there are problems with such theories, but the point is that causality and the directionality of time are far from being straightforward principles.

Moreover, even if we agree to exclude infinite speeds, i.e., that the composition of any two finite speeds must yield a finite speed, we haven't really accomplished anything, because the Galilean composition law has this same property. Every real number is finite, but it does *not* follow that there must be some finite upper bound on the real numbers. More fundamentally, it's important to recognize that the Minkowski structure of spacetime doesn't, by itself, automatically rule out speeds above the characteristic speed c (nor does it imply temporal asymmetry).

Strictly speaking, a separate assumption is required to rule out "tachyons". Thus, we can't really say that Minkowskian spacetime is prima facie any more consistent with causality than is Galilean spacetime.

A more persuasive argument for a finite upper bound on speeds can be based on the idea of locality, as mentioned in our review of the shortcomings of the Galilean transformation rule. If the spatial ordering of events is to have any absolute significance, in spite of the fact that distance can be transformed away by motion, it seems that there must be some definite limit on speeds. Also, the continuity and identity of objects from one instant to the next (ignoring the lessons of quantum mechanics) is most intelligible in the context of a unified spacetime manifold with a definite non-singular connection, which implies a finite upper bound on speeds. This is in the spirit of Minkowski's 1908 lecture in which he urged the greater "mathematical intelligibility" of the Lorentzian group as opposed to the Galilean group of transformations.

For a typical derivation of the Lorentz transformation in this axiomatic spirit, we may begin with the basic Galilean programme of seeking to identify coordinate systems with respect to which physical phenomena are optimally simple.

We have the fundamental principle that for any material object in any state of motion there exists a system of space and time coordinates with respect to which the object is instantaneously at rest and Newton's laws of inertial motion hold good (at least quasi-statically).

Such a system is called an inertial rest frame coordinate system of the object. Let x, t denote inertial rest frame coordinates of one object, and let x', t' denote inertial rest frame coordinates of another object moving with a speed v in the positive x direction relative to the x, t coordinates.

How are these two coordinate systems related? We can arrange for the origins of the coordinate systems to coincide. Also, since these coordinate systems are defined such that an object in uniform motion with respect to one such system must be in uniform motion with respect to all such systems, and such that inertia isotropic, it follows that they must be linearly related by the general form $x' = Ax + Bt$ and $t' = Cx + Dt$, where A, B, C, D are constants for a given value of v. The differential form of these equations is $dx' = Adx + Bdt$ and $dt' = Cdx + Ddt$.

Now, since the second object is stationary at the origin of the x', t' coordinates, it's position is always $x' = 0$, so the first transformation equation gives $0 = Adx + Bdt$, which implies $dx/dt = -B/A = v$ and hence $B = -Av$. Also, if we solve the two transformation equations for x and t we get $(AD-BC)x = Dx' - Bt'$, $(AD-BC)t = -Cx' + A$.

Since the first object is moving with velocity $-v$ relative to the x', t' coordinates we have $-v = dx'/dt' = B/D$, which implies

$B = -Dv$ and hence $A = D$. Furthermore, reciprocity demands that the determinant $AD - BC = A^2 + vAC$ of the transformation must equal unity, so we have $C = (1-A^2)/(vA)$. Combining all these facts, a linear, reciprocal, unitary transformation from one system of inertial coordinates to another must be of the form

$$x' = A(x - vt) \quad t' = A\left(t - vx\left[\frac{A^2 - 1}{v^2A^2}\right]\right)$$

It only remains to determine the value of A (as a function of v), which we can do by fixing the quantity in the square brackets. Letting k denote this quantity for a given v, the transformation can be written in the form

$$x' = \frac{x - vt}{\sqrt{1 - kv^2}} \quad t' = \frac{t - kvx}{\sqrt{1 - kv^2}}$$

Any two inertial coordinate systems must be related by a transformation of this form, where v is the mutual speed between them. Also, note that

$$\frac{x'}{t'} = \frac{x - vt}{t - kvx}$$

Given three systems of inertial coordinates with the mutual speed v between the first two and u between the second two, the transformation from the first to the third is the composition of transformations with parameters k_v and k_u. Letting x", t" denote the third system of coordinates, we have by direct substitution

$$\frac{x''}{t''} = \frac{x - \left(\dfrac{u + v}{1 + k_v uv}\right)t}{\left(\dfrac{1 + k_u uv}{1 + k_v uv}\right)t - \left(\dfrac{k_u u + k_v v}{1 + k_v uv}\right)x}$$

The coefficient of t in the denominator of the right side must be unity, so we have $k_u = k_v$, and therefore k is a constant for all v, with units of an inverse squared speed. Also, the coefficient of t in the numerator must be the mutual speed between the first and third coordinate systems. Thus, letting w denote this speed, we have;

$$W - \frac{u + v}{1 + kuv}$$

It's easy to show that this is the necessary and sufficient condition for the composite transformation to have the required form. Now, if the value of the constant k is non-zero, we can normalize its magnitude by a suitable choice of space and time units, so that the only three fundamentally distinct possibilities to consider are k = -1, 0, and +1. Setting k = 0 gives the familiar Galilean transformation x' = x – vt, t' = t.

This is highly asymmetrical between the time and space parameters, in the sense that it makes the transformed space parameter a function of both the space coordinate and the time coordinate of the original system, whereas the transformed time coordinate is dependent only on the time coordinate of the original system. Alternatively, for the case k = -1 we have the transformation

$$x' = \frac{x - vt}{\sqrt{1 + v^2}} \qquad t' = \frac{t - vx}{\sqrt{1 + v^2}}$$

Letting θ denote the angle that the line from the origin to the point (x, t) makes with the t axis, then tan(θ) = v = dx/dt, and we have the trigonometric identities $\cos(\theta) = 1/(1+v^2)^{1/2}$ and $\sin(\theta) = v/(1+v^2)^{1/2}$. Therefore, this transformation can be written in the form

$$x' = \cos(\theta)x - \sin(\theta)t \qquad t' = \sin(\theta)x + \cos(\theta)t$$

which is just a Euclidean rotation in the xt plane. Under this transformation the quantity $(dx)^2 + (dt)^2 = (dx')^2 + (dt')^2$ is invariant. This transformation is clearly *too* symmetrical between x and t, because know from experience that we cannot turn around in time as easily as we can turn around in space. The only remaining alternative is to set k = 1, which gives the transformation

$$x' = \frac{x - vt}{\sqrt{1 - v^2}} \qquad t' = \frac{t - vx}{\sqrt{1 - v^2}}$$

Although perfectly symmetrical, this maintains the absolute distinction between spatial and temporal intervals. This can be parameterized as a hyperbolic rotation

$$x' = \cosh(\theta)x - \sinh(\theta)t \qquad t' = \sinh(\theta)x + \cosh(\theta)t$$

and we have the invariant quantity $(dx)^2 - (dt)^2 = (dx')^2 - (dt')^2$ for any given interval.

It's hardly surprising that this transformation, rather than either the Galilean transformation or the Euclidean transformation, gives the actual relationship between space and time coordinate systems with respect to which inertia is directionally symmetrical and inertial motion is linear.

From purely formal considerations we can see that the Galilean transformation, given by setting $k = 0$, is incomplete and has no spacetime invariant, whereas the Euclidean transformation, given by setting $k = -1$, makes no distinction at all between space and time.

Only the Lorentzian transformation, given by setting $k = 1$, has completely satisfactory properties from an abstract point of view, which is presumably why Minkowski referred to it as "more intelligible".

As plausible as such arguments may be, they don't amount to a logical deduction, and one is left with the impression that we have not succeeded in identifying any fundamental principle or symmetry that uniquely selects Lorentzian spacetime rather than Galilean space and time.

Accordingly, most writers on the subject have concluded (reluctantly) that Einstein's light speed postulate, or something like it, is indispensable for deriving special relativity, and that we can be persuaded to adopt such a postulate only by empirical facts.

Indeed, later in the same paper where Minkowski exercised his staircase wit, he admitted that "the impulse and true motivation for assuming the group G_c came from the fact that the differential equation for the propagation of light [i.e., the wave equation] in empty space possesses the group G_c", and he referred back to Voigt's 1887 paper.

Nevertheless, it's still interesting to explore the various rational "intelligibility" arguments that can be put forward as to why space and time must be Minkowskian. A typical approach is to begin with three speeds u, v, w representing the pairwise speeds between three co-linear particles, and to seek a composition law of the form $Q(u, v, w) = 0$ relating these speeds.

It's easy to make the case that it should be possible to uniquely solve this function explicitly for any of the speeds in terms of the other two, which implies that Q must be linear in all three of its arguments. The most general linear function of three variables is

$$Q(u, v, w) = Auvw + Buv + Cuw + Dvw + Eu + Fv + Gw + H$$

where A, B, ...H are constants. Treating the speeds symmetrically requires $B = C = D$ and $E = F = G$.

Also, if any two of the speeds is 0 we require the third speed to be 0 (transitivity), so we have $H = 0$. Also, if any one of the speeds, say u, is 0, then we require $v = -w$ (reciprocity), but with $u = 0$ and $v = -w$ the formula reduces to $-Dv^2 + Fv - Gv = 0$, and since $F = G$ (= E) this is just $Dv^2 = 0$, so it follows that $B = C = D = 0$.

Hence the most general function that satisfies our requirements of linearity, 3-way symmetry, transitivity, and reciprocity is $Q(u, v, w) = Auvw + E(u+v+w) = 0$. It's clear that E must be non-zero (since otherwise general reciprocity would not be imposed when any one of the variables vanished), so we can divide this function by E, and let k denote A/E, to give

$$k\,uvw + u + v + w = 0$$

We see that this k is the same. As before, the only three distinct cases are $k = -1$, 0, and +1. If $k = 0$ we have the Galilean composition law, and if $k = 1$ we have the Einsteinian composition law. How are we to decide? In the next section we consider the problem from a slightly different perspective, and focus on a unique symmetry that arises only with $k = 1$.

Chapter 3

Symmetry of Relativity

Maxwell's equations are invariant under Lorentz transformations, as well as translations and spatial rotations. Together these transformations comprise the Poincare group. Of course, Maxwell's equations are also invariant under spatial and temporal reflections, but it is often overlooked that in addition to all these linear transformations, Maxwell's equations possess still another symmetry, namely, the symmetry of spacetime *inversion*.

In a sense, an inversion is a kind of reflection about a surface in spacetime, analogous to inversions about circles in projective geometry, the only difference being that the Minkowski interval is used instead of the Euclidean line element.

Consider two events E_1 and E_2 that are null-separated from each other, meaning that the absolute Minkowski interval between them is zero in terms of an inertial coordinate system x, y, z, t. Let s_1 and s_2 denote the absolute intervals from the origin to these two events (respectively).

Under an inversion of the coordinate system about the surface at an absolute interval R from the origin (which may be chosen arbitrarily), each event located on a given ray through the origin is moved to another point on that ray such that its absolute interval from the origin is changed from s to R^2/s. Thus the hyperbolic surfaces outside of R are mapped to surfaces inside R, and vice versa. To prove that two events originally separated by a null Minkowski interval are still null-separated after the coordinates have been inverted, note that the ray from the origin to the event E_j can be characterized by constants α_j, β_j, γ_j defined by

$$x_j - \alpha_j t_j \quad y_j - \beta_j t_j \quad z_j - \gamma_j t_j$$

In terms of these parameters the magnitude of the interval from the origin to E_j can be written as

$$s_j = t_j\sqrt{\alpha_2^2 + \beta_2^2 + \gamma_2^2 - 1}$$

The squared interval between E_1 and E_2 can then be expressed as

$$s_{12}^2 = (x_2 - x_1)^2 + (y_2 - y_1)^2 + (z_2 - z_1)^2 - (t_2 - t_1)^2$$

$$= (\alpha_2^2 + \beta_2^2 + \gamma_2^2 - 1)t_2^2 + (\alpha_1^2 + \beta_1^2 + \gamma_1^2 - 1)t_1^2$$

$$-2(\alpha_1\alpha_2 + \beta_1\beta_2 + \gamma_1\gamma_2 - 1)t_1t_2$$

$$= s_1^2 + s_1^2 - 2K_{12}s_1s_2$$

where

$$K_{12} = \frac{(\alpha_1\alpha_2 + \beta_1\beta_2 + \gamma_1\gamma_2 - 1)}{\sqrt{(\alpha_1^2 + \beta_1^2 + \gamma_1^2 - 1)(\alpha_2^2 + \beta_2^2 + \gamma_2^2 - 1)}}$$

Since inversion leaves each event on its respective ray, the value of K_{12} for the inverted coordinates is the same as for the original coordinates, so the only effect on the Minkowski interval between E_1 and E_2 is to replace s_1 and s_2 with R^2/s_1 and R^2/s_2 respectively. Therefore, the squared Minkowski interval between the two events in terms of the inverted coordinates is

$$(s_{12}{}^t)^2 = \frac{R^2}{s_1^2} + \frac{R^2}{s_2^2} - 2K_{12}\frac{R^2}{s_1s_2} = \frac{R^2}{s_1s_2}(s_1^2 + s_2^2 - 2K_{12}s_1s_2)$$

The quantity in parentheses on the right side is just the original squared interval, so if the interval was zero in terms of the original coordinates, it is zero in terms of the inverted coordinates. Thus inversion of a system of inertial coordinates yields a system of coordinates in which all the null intervals are preserved.

It was shown in 1910 by Bateman and (independently) Cunningham that this is the necessary and sufficient condition for Maxwell's equations to be invariant. Incidentally, Einstein was dismissive of this invariance when Felix Klein asked him about it. He wrote I am convinced that the covariance of Maxwell's formulas under transformation according to reciprocal radii can have no deeper significance; although this transformation retains the form of the equations, it does not uphold the correlation between coordinates and the measurement results from measuring rods and clocks.

Einstein was similarly dismissive of Minkowski's "formal approach" to spacetime at first, but later came to appreciate the profound significance of it. In any case, it's interesting to note that straight lines in inertial coordinate systems map to straight or hyperbolic paths under inversion. This partly accounts for the fact that, according to the Lorentz-Dirac equations of classical electrodynamics, perfect hyperbolic motion *is* inertial motion, in the sense that there are free-body solutions describing particles in hyperbolic motion, and a charged particle in hyperbolic motion does not radiate.

It's also interesting that the relativistic formula for composition of two speeds is invariant under inversion of the arguments about the speed c, i.e., replacing each speed v with c^2/v.

Letting f(u, v) denote the composition of the (co-linear) speeds u and v, and choosing units so that c = 1, we can impose the three requirements

$$0 = f(u,-u) \qquad u = f(0,u) = f(u,0) \qquad f(u,v) = f(1/u,1/v)$$

The first two requirements are satisfied by both the Galilean and the Lorentzian composition formulas, but the third requirement is not satisfied by the Galilean formula, because that gives

$$f(u,v) = u + f \qquad f(1/u,1/v) = \frac{1}{u} + \frac{1}{v} = \frac{u+v}{uv}$$

However, somewhat surprisingly, the relativistic composition function gives

$$f(u,v) = \frac{u+v}{1+uv} \qquad f(1/u,1/v) = \frac{\frac{1}{u}+\frac{1}{v}}{\frac{1}{u}\frac{1}{v}+1} = \frac{v+u}{1+uv}$$

so it does comply with all three requirements. This singles out the composition law with k = 1 from the previous chapter. As indicated by Einstein's reply to Klein, the physical significance of such inversion symmetries is obscure, and we should also note that the spacetime inversion is not equivalent to the speed inversion, although they are formally very similar. To clarify how this symmetry arises in the relativistic context, recall that we had derived at the end of the previous chapter the relation

$$k\,uvw + u + v + w = 0$$

where $u = v_{12}$, $v = v_{23}$, and $w = v_{31}$. The symbol v_{ij} signifies the speed of the ith particle in terms of the inertial rest frame coordinates of the jth particle. With $k = 0$ this corresponds to the Galilean speed composition formula, which clearly is not invariant under inversion of any or all of the speeds. For any non-zero value of k, equation can be re-written in the form

$$\left(1+u\sqrt{k}\right)\left(1+v\sqrt{k}\right)\left(1+w\sqrt{k}\right)=\left(1-u\sqrt{k}\right)\left(1-v\sqrt{k}\right)\left(1-w\sqrt{k}\right)$$

Squaring both sides of this equation gives the equality

$$\left(1+u\sqrt{k}\right)^2\left(1+v\sqrt{k}\right)^2\left(1+w\sqrt{k}\right)^2$$
$$=\left(1-u\sqrt{k}\right)^2\left(1-v\sqrt{k}\right)^2\left(1-w\sqrt{k}\right)^2$$

If we replace each speed with its inversion in this formula, and then multiply through by $(uvw)^2 / k^3$ we get

$$\left(1+\frac{u}{\sqrt{k}}\right)^2\left(1+\frac{v}{\sqrt{k}}\right)^2\left(1+\frac{w}{\sqrt{k}}\right)^2=\left(1-\frac{u}{\sqrt{k}}\right)^2\left(1-\frac{v}{\sqrt{k}}\right)^2\left(1-\frac{w}{\sqrt{k}}\right)^2$$

which is equivalent to the preceding formula if and only if

$$\sqrt{k}=\pm\frac{1}{\sqrt{k}}$$

Hence the speed composition formula is invariant under inversion if $k = \pm 1$. The case $k = -1$ is equivalent to the case $k = +1$ if each speed is taken to be imaginary (corresponding to the use of an imaginary time axis), so without loss of generality we can choose $k = +1$ with real speeds. There remains, however, the ambiguity introduced by squaring both sides of equation, suppressing the signs of the factors.

Equation itself, without squaring, is invariant under inversion of any two of the speeds, but the inversion of all three speeds changes the sign of the right side. Thus by squaring both sides we make it consistent with either of the two complementary relations

$$uvw + u + v + w = 0 \qquad 1 + uv + uw + vw = 0$$

The left hand relation is invariant under inversion of any two of the speeds, whereas the right hand relation is invariant under inversion of one or all three of the speeds. The question, then, is why the first formula applies rather than the second. To answer this, we should first point out that, despite the formal symmetry of the quantities u, v, w in these equations, they are not conceptually symmetrical. Two of the quantities are implicitly defined in terms of one inertial coordinate system, and the third quantity is defined in terms of a different inertial coordinate system.

In general, there are nine conceptually distinct speeds for three co-linear particles in terms of the three rest frame coordinate systems, namely

$$\begin{matrix} v_{11} & v_{12} & v_{13} \\ v_{21} & v_{22} & v_{23} \\ v_{31} & v_{32} & v_{33} \end{matrix}$$

where v_{ij} is the speed of the ith particle in terms of the inertial rest frame coordinates of the jth particle. By definition we have $v_{ii} = 0$ and by reciprocity we have $v_{ij} = -v_{ji}$, so the speeds comprise an anti-symmetric array.

Thus, although the three speeds v_{12}, v_{23}, v_{31} are nominally defined in terms of three different systems of coordinates, any two of them can be expressed in terms of a single coordinate system by invoking the reciprocity relation.

For example, the three quantities v_{12}, v_{23}, v_{31} can be expressed in the form v_{12}, $-v_{32}$, v_{31}, which signifies that the first two speeds are both defined in terms of the rest frame coordinates of frame 2. However, the remaining speed does not have a direct expression in terms of that frame, so a composition formula is needed to relate all three quantities. We've seen that the relativistic composition formula yields the same value for the third speed regardless of whether we use the two other speeds or their reciprocals.

To more clearly exhibit the peculiar 2+1 symmetry of this velocity composition law, note that it can be expressed in multiplicative form as

Needless to say, the above discussion is just a rough sketch, intended to show only the outline of an argument. It seems likely that Minkowski was influenced by Klein's Erlanger programme, which sought to interpret various kinds of geometry in terms of the invariants under a specific group of transformations.

It is certainly true that we are led toward the Lorentz transformations as soon as we consider the group of velocity transformations and attempt to identify a physically meaningful invariant corresponding to these transformations. However, the preceding discussion glossed over several important considerations, and contains several unstated assumptions. In the following, we will examine Minkowski's argument in more detail, paying special attention to the physical significance of each assertion along the way, and elaborating more fully the rational basis for concluding that there must be a definite relationship between the measures of space and time.

For any system of mutually orthogonal spatial coordinates x, y, z, let the positions of the two ends of a given spatially extended physical entity be denoted by x_1, y_1, z_1 and x_2, y_2, z_2, and let L^2 denote the sum of the squares of the component differences. In other words

$$s^2 = (x_2 - x_1)^2 + (y_2 - y_1)^2 + (z_2 - z_1)^2$$

Experience teaches us that, for a large class of physical entities ("solids"), we can shift and/or re-orient the entity (relative to the system of coordinates), changing the individual components, but the sum of the squares of the component differences remains unchanged. The invariance of this quantity under re-orientations is called spatial isotropy.

It's worth emphasizing that the invariance of s^2 under these operations applies only if the x, y, and z coordinates are mutually orthogonal. The spatial isotropy of physical entities implies a non-trivial unification of orthogonal measures. Strictly speaking, each of the three terms on the right side should be multiplied by a coefficient whose units are the squared units of s divided by the squared units of x, y, or z respectively. In writing the equation without coefficients, we have tacitly chosen units of measure for x, y, and z such that the respective coefficients are 1. In addition, we tacitly assumed the spatial coordinates of the two ends of the

$$\left(\frac{1+v_{12}}{1-v_{12}}\right)\left(\frac{1+v_{23}}{1-v_{23}}\right)\left(\frac{1+v_{31}}{1-v_{31}}\right)=1$$

where v_{ij} denotes the speed of object j with respect to object i. Clearly if we replace any two of the speeds with their reciprocals, the relation remains unchanged. On the other hand, if we replace just one or all three of the speeds with their reciprocals, their product is still unity, but the sign is negated. Thus, one way of expressing the full symmetry of this relation would be to square both sides, giving the result

$$\left(\frac{1+v_{12}}{1-v_{12}}\right)^2\left(\frac{1+v_{23}}{1-v_{23}}\right)^2\left(\frac{1+v_{31}}{1-v_{31}}\right)^2=1$$

which is completely invariant under any replacement of one or more speeds with their respective reciprocals. Naturally we can extend the product of factors of the form $(1+v_{ij})/(1-v_{ij})$ to any cyclical sequence of relative speeds between any number of co-linear points.

It's interesting to note the progression of relations between the speeds involving one, two, and three particles. The relativity of position is expressed by the identity

$$v_{ii}=0$$

for any one particle, and the relativity of velocity can be expressed by the skew symmetry

$$v_{ij}+v_{ji}=0$$

for any two systems particles. (This was referred to earlier as the reciprocity condition $v_{ij} = -v_{ji}$). The next step is to consider the cyclic sum involving three particles and their respective inertial rest frame coordinate systems. This is the key relation, because all higher-order relations can be reduced to this. If acceleration were relative (like position and velocity), we would expect the cyclic symmetry $v_{ij} + v_{jk} + v_{ki} = 0$, which is a linear function of all three components. Indeed, this is the Galilean composition formula.

However, since acceleration is absolute, it's to be expected that the actual relation is non-linear in each of the three components. So, instead of vanishing, we need the right side of this sum to be a symmetric function of the terms. The only other odd elementary symmetric function of three quantities is the product of all three, so we're led (again) to the relation

$$v_{ij} + v_{jk} + v_{ki} = -v_{ij} v_{jk} v_{ki}$$

which can be regarded as the law of inertia. Since there is only one odd elementary symmetric function of one variable, and likewise for two variables, the case of three variables is the first for which there exists a non-tautological expression of this form.

We may also note a formal correspondence with De Morgan's law for logical statements. Letting sums denote logical ORs (unions), products denote logical ANDs (intersections), and overbars denote logical negation, De Morgan's law states that

$$\overline{X + Y + Z} = \overline{X}\ \overline{Y}\ \overline{Z}$$

for any three logical variables X, Y, Z. Now, using the skew symmetry property, we can "negate" each velocity on the right hand side of the previous expression to give

$$v_{ij} + v_{jk} + v_{ki} = v_{ji} v_{kj} v_{ik}$$

From this standpoint the right hand side is analogous to the "logical negation" of the left hand side, which makes the relation analogous to setting the quantity equal to zero.

The justification for regarding this relation as the source of inertia becomes, which describes how the relativistic composition law for velocities accounts for the increasing inertia of an accelerating object. This leads to the view that inertia itself is, in some sense, a consequence of the non-linearity of velocity compositions.

Given the composition law u' = (u+v)/(1+uv) for co-linear speeds, what can we say about the transformation of the coordinates x and t themselves under the action of the velocity v? The composition law can be written in the form vuu'+u'−u = v, which has a natural factorization if we multiply through by v and subtract 1 from both sides, giving

$$(1 + vu)(1 - uv') = 1 - v^2$$

If u and u' are taken to be the spatio-temporal ratios x/t and x'/t', the above relation can be written in the form

$$\left(1+v\frac{x}{t}\right)\left(1-v\frac{x'}{t'}\right)=1-v^2$$

On the other hand, remembering that we can insert the reciprocals of any two of the quantities u, u', v without disturbing the equality, we can take u and u' to be the temporal-spatial ratios t/x and t'/x' to give

$$\left(1+v\frac{t}{x}\right)\left(1-v\frac{t'}{x'}\right)=1-v^2$$

These last two equations immediately give

$$t't=\frac{(t+vx)(t'+vx')}{1-v^2} \qquad x'x=\frac{(x+vt)(x'+vt')}{1-v^2}$$

Treating the primed and unprimed frames equivalently, and recalling that v' = –v, we see that a perfectly symmetrical factorization, so we exploit this factorization to give the transformation equations

$$t'=\frac{t+vx}{\sqrt{1-v^2}} \qquad x'=\frac{x+vt}{\sqrt{1-v^2}}$$

$$t=\frac{t'+vx'}{\sqrt{1-v^2}} \qquad x=\frac{x'+vt'}{\sqrt{1-v^2}}$$

These are the Lorentz transformations for velocity v in the x direction. The y and z coordinates are unaffected, so we have y' = y and z' = z.

From this it follows that the quantity $t^2 - x^2 - y^2 - z^2$ is invariant under a general Lorentz transformation, so we have arrived at the full Minkowski spacetime metric.

Now, to determine the full velocity composition law for two systems of aligned coordinates k and K, the latter moving in the positive x direction with velocity v relative to the former, we can without loss of generality make the origins of the two systems both coincide with a point P_0 on the subject worldline, and let P_1 denote a subsequent point on that worldline with k system coordinates dt, dx, dy, dz.

By definition the velocity components of that worldline with respect to k are u_x = dx/dt, u_y = dy/dt, and u_z = dz/dt. The coordinates of P_1 with respect to the K system are given by the Lorentz transformation for a simple boost v in the x direction:

$$dX = \frac{dx - vdt}{\gamma} \quad dY = dy \quad dZ = dz \quad dT = \frac{dt - vdx}{\gamma}$$

where $\gamma = \sqrt{1 - v^2}$. Therefore, the velocity components of the worldline with respect to the K system are

$$U_X = \frac{dx - vdt}{dt - vdx} = \frac{u_x - v}{1 - vu_x}$$

$$U_Y = \frac{\gamma dy}{(dt - vdx)} = \frac{\gamma\, u_y}{(1 - vu_x)}$$

$$U_Z = \frac{\gamma dz}{(dt - vdx)} = \frac{\gamma\, u_z}{(1 - vu_x)}$$

NULL COORDINATES

Initially the special theory of relativity was regarded as just a particularly simple and elegant interpretation of Lorentz's ether theory, but it soon became clear that there is a profound heuristic difference between the two theories, most evident when we consider the *singularity* implicit in the Lorentz transformation $x' = \gamma(x-vt)$, $t' = \gamma(t-vx)$, where $\gamma = 1/(1-v^2)^{1/2}$. As v approaches arbitrarily close to 1, the factor γ goes to infinity.

If these relations are strictly valid (locally), as all our observations and experiements suggest, then according to Lorentz's view all configurations of objects moving through the absolute ether must be capable of *infinite* spatial "contractions" and temporal "dilations", without the slightest distortion.

This is clearly unrealistic. Hence the only plausible justification for the Lorentzian view is a belief that the Lorentz transformation equations are *not* strictly valid, i.e., that they must break down at some point. Indeed, this was Lorentz's ultimate justification, as he held to the possibility that absolute speed might, after all, make some difference to the intrinsic relations between physical entities.

However, one hundred years after Lorentz's time, there still is no evidence to support his suspicion. To the contrary, all the tremendous advances of the last century in testing the Lorentz transformation "to the nth degree" have consistently confirmed it's *exact* validity. At some point a reasonable person must ask himself "What if the Lorentz transformation really is *exactly* correct?" This is a possibility that a neo-etherist cannot permit himself to contemplate - because the absolute physical singularity along light-like intervals implied by the Lorentz transformation is plainly incompatible with any realistic ether - but it is precisely what special relativity *requires* us to consider, and this ultimately leads to a completely new and more powerful view of causality.

The singularity of the Lorentz transformation is most clearly expressed in terms of the underlying Minkowski pseudo-metric. Recall that the invariant space time interval $d\tau$ between the events (t, x) and (t+dt, x+dx) is given by

$$(d\tau)^2 = (d\tau)^2 - (dx)^2$$

where t and x are any set of inertial coordinates. This is called a pseudo-metric rather than a metric because, unlike a true metric, it doesn't satisfy the triangle inequality, and the interval between distinct points can be zero.

This occurs for any interval such that dt = dx, in which case the invariant interval dt is literally zero. Arguably, it is only in the context of Minkowski spacetime, with its null connections between distinct events, that phenomena involving quantum entanglement can be rationalized. Pictorially, the locus of points whose squared distance from the origin is ±1 consists of the two hyperbolas labeled +1 and -1 in the figure below:

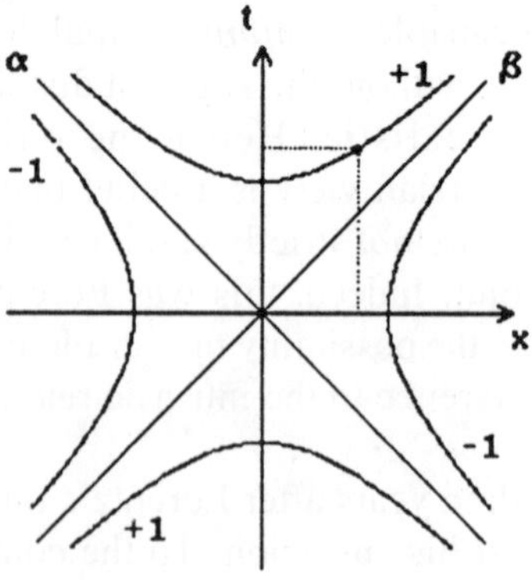

Space-Time Coordinates

The diagonal axes denoted by α and β represents the paths of light through the origin, and the magnitude of the squared spacetime interval along these axes is 0, i.e., the metric is degenerate along those lines. This is all expressed in terms of conventional space and time coordinates, but it's also possible to define the spacetime separations between events in terms of null coordinates along the light-line axes. Conceptually, we rotate the above figure by 45 degrees, and regard the α and β lines as our coordinate axes, as shown below:

In terms of a linear parameterization (α, β) of these "null coordinates" the locus of points at a squared "distance" $(d\tau)^2$ from the origin is an orthogonal hyperbola satisfying the equation

$$(d\tau)^2 = (d\alpha)(d\beta)$$

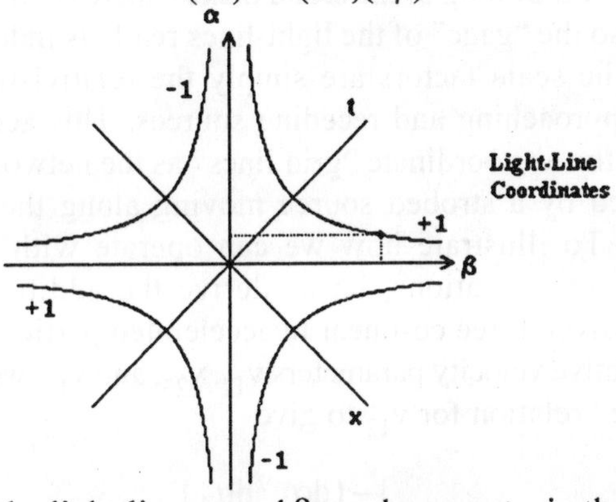

Since the light-lines α and β are degenerate, in the sense that the absolute spacetime intervals along those lines vanish, the absolute velocity of a worldline, given by the "slope" $d\beta/d\alpha = 0/0$, is strictly undefined.

This indeterminacy, arising from the singular null intervals in spacetime, is at the heart of special relativity, allowing for infinitely many different scalings of the light-line coordinates. In particular, it is natural to define the rest frame coordinates α, β of any worldline in such a way that $d\alpha/d\beta = 1$.

This expresses the principle of relativity, and also entails Einstein's second principle, i.e., that the (local) velocity of light with respect to the natural measures of space and time for any worldline is unity.

The relationship between the natural null coordinates of any two worldlines is then expressed by the requirement that, for any given interval $d\tau$, the components $d\alpha$, $d\beta$ with respect to one frame are related to the components $d\alpha'$, $d\beta'$ with respect to another frame according to the equation $(d\alpha)(d\beta) = (d\alpha')(d\beta')$. It follows that the scale factors of any two frames S_i and S_j are related according to

$$d\alpha_i = d\alpha_j\sqrt{\frac{1+v_{ij}}{1-v_{ij}}} \qquad d\beta_i = d\beta_j\sqrt{\frac{1-v_{ij}}{1+v_{ij}}}$$

where v_{ij} is the usual velocity parameter (in units such that $c = 1$) of the origin of S_j with respect to S_i. Notice there is no absolute constraint on the scaling of the α and β axes, there is only a relative constraint, so the "gage" of the light-lines really is indeterminate.

Also, the scale factors are simply the relativistic Doppler shifts for approaching and receding sources. This accords with the view of the ab coordinate "grid lines" as the network of light-lines emitted by a strobed source moving along the reference world-line. To illustrate how we can operate with these null coordinate scale relations, let us derive the addition rule for velocities. Given three co-linear unaccelerated particles with the pairwise relative velocity parameters v_{12}, v_{23}, and v_{13}, we can solve the "α scale" relation for v_{13} to give

$$v_{13} = \frac{1-(d\alpha_1/d\alpha_3)^2}{1+(d\alpha_1/d\alpha_3)^2}$$

We also have

$$\left(\frac{d\alpha_1}{d\alpha_2}\right)^2 = \frac{1-v_{12}}{1+v_{12}} \qquad \left(\frac{d\alpha_2}{d\alpha_3}\right)^2 = \frac{1-v_{23}}{1+v_{23}}$$

Multiplying these together gives an expression for $d\alpha_1/d\alpha_3$, which can be substituted to give the expected result

$$v_{13} = \frac{v_{12}+v_{23}}{1+v_{12}v_{23}}$$

Interestingly, although neither the velocity parameter v nor the quantity (1+v)/(1−v) is additive, it's easy to see that the parameter ln[(1+v)/(1−v)] is additive. In fact, this parameter corresponds to the arc length of the "dτ = constant" hyperbola connecting the two world lines at unit distances from their intersection, as shown by integrating the differential distance along that curve

$$ds = \sqrt{(dx)^2 - (dt)^2} = \sqrt{1 - (dx/dx)^2}\,dx$$

Since the equation of the hyperbola for dτ = 1 is $1 = dt^2 - dx^2$ we have

$$\frac{dt}{dx} = \frac{x}{\sqrt{1+x^2}}$$

Substituting this into the previous expression and performing the integration gives

$$ds = \ln\left(dx + \sqrt{1+dx^2}\right) = \ln(dx + dt)$$

Recalling that $d\tau^2 = dt^2 - dx^2$, we have $dt + dx = d\tau^2 / (dt - dx)$, so the quantity dx + dt can be written as

$$dt + dx = d\tau\sqrt{\frac{dt + dx}{dt - dx}} = d\tau\sqrt{\frac{1+v}{1-v}}$$

Hence the absolute arc length along the dτ = 1 surface between two world lines that intersect at the origin with a mutual velocity v is

$$s = \frac{1}{2}\ln\left(\frac{1+v}{1-v}\right)$$

Naturally the additivity of this logarithmic form implies that the argument is a multiplicative measure of mutual speeds. The absolute interval between the intersection points of the two worldlines with the dτ = 1 hyperbola is

$$s = \left(\frac{1+v}{1-v}\right)^{1/4} - \left(\frac{1-v}{1+v}\right)^{1/4}$$

One strength of the conventional pseudo-metrical formalism is that (t, x) coordinates easily generalize to (t, x, y, z) coordinates, and the invariant interval generalizes to

$$(d\tau)^2 = (dt)^2 - (dx)^2 - (dy)^2 - (dz)^2$$

The generalization of the null (lightlike) coordinates and corresponding invariant is not as algebraically straightforward, but it conveys some interesting aspects of the spacetime structure. Intuitively, an observer can conceive of the absolute interval between himself and some distant future event P by first establishing a scale of radial measure outward on his forward light cone in all directions, and then for each direction evaluate the param eterized *null measure* along the light cone to the point of intersection with the backward null cone of P.

This will assign, to each direction in space, a parameterized distance from the observer to the backward light cone of P, and there will be (in flat spacetime) two distinguished directions, along which the null measure is maximum or minimum.

These are the *principle directions* for the interval from the observer to E, and the product of the null measures in these directions is invariant.

In other words, if a second observer, momentarily coincident with the first but with some relative velocity, determines the null measures along the principle directions to the backward light cone of E, with respect to his own natural parameterization, the product will be the same as found by the first observer.

It's often convenient to take the interval to the point P as the time axis of inertial coordinates t, x, y, z, so the eigenvectors of the null cone intersections become singular, and we can simply define the null coordinates u = t + r, v = t – r, where $r = (x^2+y^2+z^2)^{1/2}$. From this we have t = (u+r)/2 and r = (u–v)/2 along with the corresponding differentials dt = (du+dv)/2 and dr = (du–dv)/2. Making these substitutions into the usual Minkowski metric in terms of polar coordinates

$$(d\tau)^2 = (dt)^2 - (dr)^2 - r^2 - \left[(d\theta)^2 + \sin\,(\theta)^2\,(d\phi)^2\right]$$

we have the Minkowski line element in terms of angles and null coordinates

$$(d\tau)^2 = drdv - \left(\frac{u - v}{2}\right)^2 \left[(d\theta)^2 + \sin\,(\theta)^2\,(d\phi)^2\right]$$

These coordinates are often useful, but we can establish a more generic system of null coordinates in 3+1 dimensional spacetime by arbitrarily choosing four non-parallel directions in space from an observer at O, and then the coordinates of any timelike separated event are expressed as the four null measures radially in those directions along the forward null cone of O to the backward null cone of P. This provides enough information to fully specify the interval OP.

In terms of the usual orthogonal spacetime coordinates, we specify the coordinates (T, X, Y, Z) of event P relative to the observer O at the origin in terms of the coordinates of four events I_1, I_2, I_3, I_4 on the intersection of the forward null cone of O and the backward null cone of P. If t_i, x_i, y_i, z_i denote the conventional coordinates of I_i, then we have

$$t_i^2 = x_i^2 + y_i^2 + z_i^2 \quad (T - t_i)^2 = (X - x_i)^2 + (Y - y_i)^2 + (Z - z_i)^2$$

for i = 1, 2, 3, 4. Expanding the right hand equations and canceling based on the left hand equalities, we have the system of equations

$$T^2 - X^2 - Y^2 - Z^2 = 2(Tt_i - Xx_i - Yy_i - Zz_i)$$

The left hand side of all four of these equations is the invariant squared proper time interval τ^2 from O to P, and we wish to express this in terms of just the four null measures in the four chosen directions.

For a specified set of directions in space, this information can be conveyed by the four values t_1, t_2, t_3, and t_4, since the magnitudes of the spatial components are determined by the directions of the axes and the magnitude of the corresponding t. In general we can define the direction coefficients a_{ij} such that

$$x_i = a_{i1}t_i \qquad y_i = a_{i2}t_i \qquad z_i = a_{i3}t_i \qquad 1=1,2,3,4$$

with the condition $a_{i1}^2 + a_{i2}^2 + a_{i3}^2 = 1$. Making these substitutions, the system of equations can be written in matrix form as

$$\begin{bmatrix} 1 & a_{11} & a_{12} & a_{13} \\ 1 & a_{21} & a_{22} & a_{23} \\ 1 & a_{31} & a_{32} & a_{33} \\ 1 & a_{41} & a_{42} & a_{43} \end{bmatrix} \begin{bmatrix} T \\ -X \\ -Y \\ -Z \end{bmatrix} = \frac{\tau^2}{2} \begin{bmatrix} 1/t_1 \\ 1/t_2 \\ 1/t_3 \\ 1/t_4 \end{bmatrix}$$

We can use any four directions for which the determinant of the coefficient matrix does not vanish. One natural choice is to use the vertices of a tetrahedron inscribed in a unit sphere, so that the four directions are perfectly symmetrical. We can take as the coordinates of the vertices

$$(1,0,0)(-1/3, 2\sqrt{23}, 0)(-1/3, -\sqrt{2}/3, \sqrt{2/3})$$
$$(-1/3, -\sqrt{2}/3, -\sqrt{2/3})$$

Inserting these values for the direction coefficients a_{ij}, we can solve the matrix equation for T, X, Y, and Z to give

$$T = \frac{\tau^2}{8}\left(\frac{1}{t_1}+\frac{1}{t_2}+\frac{1}{t_3}+\frac{1}{t_4}\right) X = -\frac{\tau^2}{8}\left(\frac{3}{t_1}+\frac{1}{t_2}+\frac{1}{t_3}+\frac{1}{t_4}\right)$$

$$Y = -\frac{\tau^2}{8}\sqrt{2}\left(\frac{2}{t_2}-\frac{1}{t_3}-\frac{1}{t_4}\right) \quad Z = -\frac{\tau^2}{8}\sqrt{6}\left(\frac{1}{t_3}-\frac{1}{t_4}\right)$$

Substituting into the relation $\tau^2 = T^2 - X^2 - Y^2 - Z^2$ and solving for τ^2 gives

$$\frac{16}{\tau^2} = \left(\frac{1}{t_1}+\frac{1}{t_2}+\frac{1}{t_3}+\frac{1}{t_4}\right)^2 - 3\left(\frac{1}{t_1^2}+\frac{1}{t_2^2}+\frac{1}{t_3^2}+\frac{1}{t_4^2}\right)$$

Naturally if $t_1 = t_2 = t_3 = t_4 = t$, then this gives $\tau = \pm 2t$. Also, notice that, as expected, this expression is perfectly symmetrical in the four lightlike coordinates. It's interesting that if the right hand term was absent, then τ would be simply the harmonic mean of the t_i.

More generally, in a spacetime of 1 + (D–1) dimensions, the invariant interval in terms of D perfectly symmetrical null measures $t_1, t_2, \ldots, t_D$ satisfies the equation

$$\frac{4D}{\tau^2} = \left(\sum \frac{1}{t_j}\right)^2 - (D-1)\left(\sum \frac{1}{t_j^2}\right)$$

It can be verified that with D = 2 this expression reduces to $\tau^2 = 4t_1t_2$, which agrees with our earlier hyperbolic formulation $\tau^2 = \alpha\beta$ with $\alpha = 2t_1$ and $\beta = 2t_2$. In the particular case D = 4, if we define $U = 2/\tau$ and $u_j = 1/(2t_j)$ this equation can be written in the form

$$U^2 = \sum u_j^{\,2} - \sigma^2$$

where s is the average squared difference of the individual u terms from the average, i.e.,

$$\sigma^2 = \frac{1}{4}\sum\left(u_j - \bar{u}\right)^2$$

This is the statistical variance of the u_j values. Incidentally, we've seen that the usual representation $s^2 = x^2 - t^2$ of the invariant spacetime interval is a generalization of the familiar Pythagorean "sum-of-squares" equation of a circle, whereas the interval can also be expressed in the hyperbolic form $s^2 = \alpha\beta$. This reminds us of other fundamental relations of physics that have found expression as hyperbolic relations, such as the uncertainty relations

$$(\Delta x)(\Delta p_x) \geq \frac{h}{4\pi} \qquad (\Delta E)(\Delta t) \geq \frac{h}{4\pi}$$

in quantum mechanics, where h is Planck's constant. In general if the operators A, B corresponding to two observables do not commute (i.e., if $AB - BA \neq 0$), then an uncertainty relation applies to those two observables, and they are said to be incompatible. Spatial position and momentum are maximally incompatible, as are energy and time. Such pairs of variables are called conjugates.

This naturally raises the question of whether the variables parameterizing two oppositely directed null rays in spacetime can, in some sense, be regarded as conjugates, accounting for the invariance of their product. Indeed the special theory of relativity can be interpreted in terms of a fundamental limitation on our ability to make measurements, just as can the theory of quantum mechanics.

In quantum mechanics we say that it's not possible to simultaneously measure the values of two conjugate variables such that the product of the uncertainties of those two measurements is less than $h/4\pi$. Likewise in special relativity we could say that it's not possible to measure the time difference dt between two events separated by the spatial distance dx such the ratio dt/dx of the variables is less than 1/c.

In quantum mechanics we may imagine that the particle possesses a precise position and momentum, even though we are unable to determine it due to practical limitations of our measurement techniques.

If only we have infinitely weak signal, i.e., if only h = 0, we could measure things with infinite precision. Likewise in special relativity we may imagine that there is an absolute and precise relationship between the times of two distant events, but we are prevented from determining it due to the practical limitations. If only we had an infinnitely fast signal, i.e., if only 1/c was zero, we could measure things with infinite precision. In other words, nature possesses structure and information that is inaccessible to us (hidden variables), due to the limitations of our measuring capabilities.

However, it's also possible to regard the limitations imposed by quantum mechanics ($h \neq 0$) and special relativity ($1/c \neq 0$) not as limitations of measurement, but as expressions of an actual ambiguity and "incompatibility" in the independent meanings of those variables. Einstein's central contribution to modern relativity was the idea that there is no one "true" simultaneity between spatially separate events, but rather spacetime events are only partially ordered, and the decomposition of space and time into separate variables contains an inherent ambiguity on the scale of 1/c. In other words, he rejected Lorentz's "hidden variable" approach, and insisted on treating the ambiguity in the spacetime decomposition as fundamental.

This is interesting in part because, when it came to quantum mechanics, Einstein's instinct was to continue trying to find ways of measuring the "hidden variables", and he was never comfortable with the idea that the Heisenberg uncertainty relations express a fundamental ambiguity in the decomposition of conjugate variables on the scale of h. (Late in life, as Einstein continued arguing against Bohr's notion of complementarity in quantum mechanics, one of his younger collegues said "But Professor Einstein, you yourself originated this kind of positivist reasoning about conjugate variables in the theory of space and time",

Another model suggested by the relativistic interpretation of spacetime is to conceive of space and time as two superimposed waves, combining constructively in the directions of the space and time axes, but destructively (i.e., cancelling out) along light lines. For any given inertial coordinate system x, t, we can associate with each event an angle θ defined by tan(θ) = t/x. Thus the interval from the origin to the point x, t makes an angle θ with the positive x axis, and we have t = x tan(θ), so we can express the squared magnitude of a spacelike interval as

$$s^2 = x^2 - t^2 = x^2\left[1 - \tan(\theta)^2\right] = x^2 \frac{\cos(2\theta)}{\cos(\theta)^2}$$

Multiplying through by $\cos(\theta)^2$ gives

$$s^2 \cos(\theta)^2 = x^2 \cos(2\theta)$$

Substituting $t^2 / \tan(\theta)^2$ for x^2 gives the analogous expression

$$s^2 \sin(\theta)^2 = t^2 \cos(2\theta)$$

Adding these two expressions gives the result

$$s^2 = \left(x^2 + t^2\right)\cos(2\theta)$$

Consequently the "circular" locus of events satisfying $x^2 + t^2 = r^2$ for any fixed r can be represented in polar coordinates (s, θ) by the equation

$$s^2 = r^2 \cos(2\theta)$$

which is the equation of two lemniscates, as illustrated below.

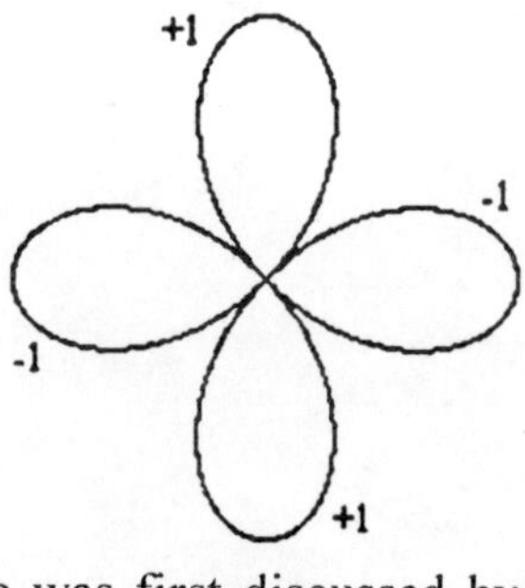

The lemniscate was first discussed by Jakob Bernoulli in 1694, as the locus of points satisfying the equation

$$x^2 + y^2 = k^2\sqrt{x^2 - y^2}$$

which is, in Bernoulli's words, "a lying eight-like figure, folded in a knot of a bundle, or of a lemniscus, a knot of a French ribbon". (The study of this curve led Fagnano, Euler, Legendre, Gauss, and others to the discovery of addition theorems for integrals, of which the relativistic velocity composition law is an example). Notice that the lemniscate is the inverse (in the sense of inversive geometry) of the hyperbola relative to the circle of radius k.

In other words, if we draw a line emanating from the origin and it strikes the lemniscate at the radius s, then it strikes the hyperbola at the radius R where $sR = k^2$.

This follows from the fact that the equation for a hyperbola in polar coordinates is $R^2 = k^2/[E^2 \cos(\theta)^2 - 1]$ where E is the eccentricity, and for an orthogonal hyperbola we have $E = \sqrt{2}$. Hence the denominator is $2\cos(\theta)^2 - 1 = \cos(2\theta)$, and the equation of the hyperbola is $R^2 = k^2/\cos(2\theta)$. Since the polar equation for the lemniscate is $s^2 = k^2\cos(2\theta)$ we have $sR = k^2$.

Chapter 4

Doppler Shift for Sound and Light

For historical reasons, some older text books present two different versions of the Doppler shift equations, one for acoustic phenomena based on traditional Newtonian kinematics, and another for optical and electromagnetic phenomena based on relativistic kinematics.

This sometimes gives the impression that relativity requires us to apply a different set of kinematical rules to the propagation of sound than to the propagation of light, but of course that is not the case. The kinematics of relativity apply uniformly to the propagation of all kinds of signals, provided we give the exact formulae.

The traditional acoustic formulas are inexact, tacitly based on Newtonian approximations, but when they are expressed exactly we find that they are perfectly consistent with the relativistic formulas.

Consider a frame of reference in which the medium of signal propagation is assumed to be at rest, and suppose an emitter and absorber are located on the x axis, with the emitter moving to the left at a speed of v_e and the absorber moving to the right, directly away from the emitter, at a speed of v_a. Let c_s denote the speed at which the signal propagates with respect to the medium. Then, according to the classical (non-relativistic) treatment, the Doppler frequency shift is

$$\nu_a = \left(\frac{1 - v_a / c_s}{1 + v_e / c_s} \right) \nu_e$$

(It's assumed here that v_a and v_e are less than c_s, because otherwise there may be shock waves and/or lack of communication between transmitter and receiver, in which case the Doppler effect does not apply). The above formula is often quoted as the Doppler effect for sound, and then another formula is given for light, suggesting that relativity arbitrarily treats sound and light signals differently. In truth, relativity has just a single formula for the Doppler shift, which applies equally to both sound and light. This formula can basically be read directly off the spacetime diagram shown below:

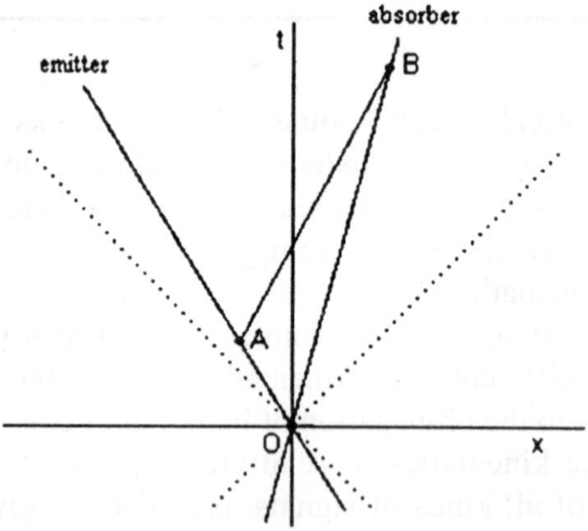

If an emitter on worldline OA turns a signal ON at event O and OFF at event A, the proper duration of the signal is the magnitude of OA, and if the signal propagates with the speed of the worldline AB, then the proper duration of the pulse for a receiver on OB will equal the magnitude of OB. Thus we have

$$c_s = \frac{x_B - x_A}{t_B - t_A} \qquad v_e = \frac{x_A}{t_B} \qquad v_a = \frac{x_B}{t_B}$$

and

$$|OA| = \sqrt{t_A{}^2 - (x_A/c)^2} = t_A\sqrt{1-(v_e/c)^2}$$

$$|OB| = \sqrt{t_B{}^2 - (x_B/c)^2} = t_B\sqrt{1-(v_a/c)^2}$$

Substituting $x_A = -v_e t_A$ and $x_B = v_a t_B$ into the equation for c_s and re-arranging terms gives

$$(t_B - t_A)c_s = v_a t_B + v_e t_A$$

from which we get

$$\frac{t_A}{t_B} = \frac{1 - v_a / c_s}{1 + v_e / c_s}$$

Substituting this into the ratio of |OA| / |OB| gives the ratio of proper times for the signal, which is the inverse of the ratio of frequencies:

$$\frac{\nu_a}{\nu_e} = \frac{|OA|}{|OB|} = \frac{1 - v_a}{1 + v_e / c_s} \sqrt{\frac{1 - (v_e / c)^2}{1 - (v_a / c)^2}}$$

Now, if v_a and v_e are both small compared to c, it's clear that the relativistic correction factor (the square root quantity) will be indistinguishable from unity, and we can simply use the leading factor, which is the classical Doppler formula for both sound and light. However, if v_a and/or v_e are fairly large (i.e., on the same order as c) we can't neglect the relativistic correction.

It may seem surprising that the formula for sound waves in a fixed medium with absolute speeds for the emitter and absorber is also applicable to light, but notice that as the signal propagation speed c_s goes to c, the above Doppler formula smoothly evolves into

$$\nu_a = \nu_e \sqrt{\left(\frac{1 - v_a / c}{1 + v_a / c}\right)\left(\frac{1 - v_e / c}{1 + v_e / c}\right)}$$

which is very nice, because we immediately recognize the quantity inside the square root as the multiplicative form of the relativistic composition law for velocities. In other words, letting u denote the composition of the speeds v_a and v_e given by the formula

$$u = \frac{v_a + v_e}{1 + v_a v_e / c_2}$$

it follows that

$$\left(\frac{1 - u / c}{1 + u / c}\right) = \left(\frac{1 - v_a / c}{1 + v_a / c}\right)\left(\frac{1 - v_e / c}{1 + v_e / c}\right).$$

Consequently, as c_s increases to c, the absolute speeds v_e and v_a of the emitter and absorber relative to the fixed medium merge into a single relative speed u between the emitter and absorber, independent of any reference to a fixed medium, and we arrive at the relativistic Doppler formula for waves propagating at c for an emitter and absorber with a relative velocity of u:

$$\nu_a = \nu_e \sqrt{\frac{1 - u/c}{1 + u/c}}$$

To clarify the relation between the classical and relativistic Doppler shift equations, recall that for a classical treatment of a wave with characteristic speed c_s in a material medium the Doppler frequency shift depends on whether the emitter or the absorber is moving relative to the fixed medium. If the absorber is stationary and the emitter is receding at a speed of v (normalized so $c_s = 1$), then the frequency shift is given by

$$\frac{\nu_a}{\nu_e} = \frac{1}{1+v} = 1 - v + v^2 + \ldots$$

whereas if the emitter is stationary and the absorber is receding the frequency shift is

$$\frac{\nu_a}{\nu_e} = 1 - v$$

To the first order these are the same, but they obviously differ significantly if v is close to 1. In contrast, the relativistic Doppler shift for light, with $c_s = c$, does not distinguish between emitter and absorber motion, but simply predicts a frequency shift equal to the geometric mean of the two classical formulas, i.e.,

$$\frac{\nu_a}{\nu_e} = \sqrt{\frac{1-v}{1+v}} = 1 - v + \frac{1}{2}v^2 - \ldots.$$

Naturally to first order this is the same as the classical Doppler formulas, but it differs from *both* of them in the second order, so we should be able to check for this difference, provided we can arrange for emitters and/or absorbers to be moving with significant speeds.

The Doppler effect *has* in fact been tested at speeds high enough to distinguish between these two formulas. The possibility of such a test, based on observing the Doppler shift for "canal rays" emitted from high-speed ions, had been considered by Stark in 1906, and Einstein published a short paper in 1907 deriving the relativistic prediction for such an experiment. However, it wasn't until 1938 that the experiment was actually performed with enough precision to discern the second order effect.

In that year, Ives and Stilwell shot hydrogen atoms down a tube, with velocities (relative to the lab) ranging from about 0.8 to 1.3 times 10^6 m/sec. As the hydrogen atoms were in flight they emitted light in all directions. Looking into the end of the tube (with the atoms coming toward them), Ives and Stilwell measured a prominent characteristic spectral line in the light coming forward from the hydrogen.

This characteristic frequency ν was Doppler shifted toward the blue by some amount $d\nu_{approach}$ because the source was approaching them.

They also placed a mirror at the opposite end of the tube, behind the hydrogen atoms, so they could look at the same light from behind, i.e., as the source was effectively moving away from them, red-shifted by some amount $d\nu_{receed}$. The following is a table of results from the original 1938 experiment for four different velocities of the hydrogen atom:

	$(10^5)(d\nu_{approach} - d\nu_{receed})\nu$		
v, 106 m/s	*Classical*	*Relativistic*	*Expreiment*
0.865	1.67	0.835	0.762
1.01	2.26	1.13	1.1
1.15	2.90	1.45	1.42
1.33	3.94	1.97	1.9

Ironically, although the results of their experiment brilliantly confirmed Einstein's prediction based on the special theory of relativity, Ives and Stillwell were not advocates of relativity, and in fact gave a completely different theoretical model to account for their experimental results and the deviation from the classical prediction.

This illustrates the fact that the results of an experiment can never uniquely identify *the* explanation. They can only split the range of available models into two groups, those that are consistent with the results and those that aren't. In this case it's clear that any model yielding the classical prediction is ruled out, while the Lorentz/Einstein model is found to be consistent with the observed results. All the above was based on the assumption that the emitter and absorber are moving relative to each other directly along their "line of sight". More generally, we can give the Doppler shift for the case when the (inertial) motions of the emitter and absorber are at any specified angles relative to the "line of sight". Without loss of generality we can assume the absorber is stationary at the origin of inertial coordinates and the emitter is moving at a speed v and at an angle ϕ relative to the direct line of sight, as illustrated below.

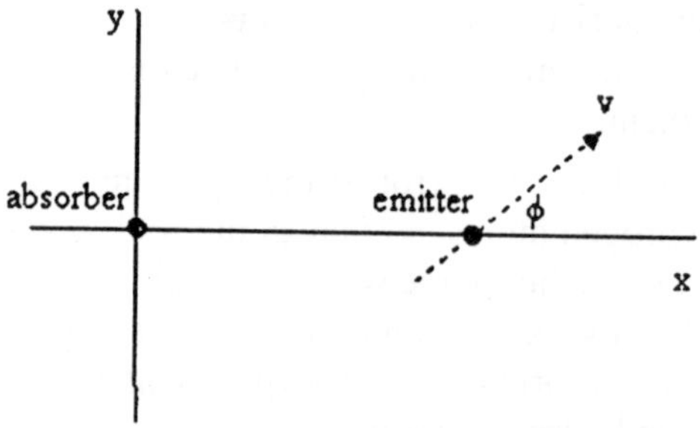

For two pulses of light emitted at coordinate times differing by Δt_e, arrival times at the receiver will differ by $\Delta t_a = (1 + v_r)\,\Delta t$ where $v_r = v \cos(\phi)$ is the radial component of the emitter's velocity. Also, the proper time interval along the emitter's worldline between the two emissions is $\Delta\tau_e = \Delta t_e\,(1 - v^2)^{1/2}$.

Therefore, since the frequency of the transmissions with respect to the emitter's rest frame is proportional to $1/\Delta\tau_e$, and the frequency of receptions with respect to the absorber's rest frame is proportional to $1/\Delta t_a$, the full frequency shift is

$$\nu_a = \frac{\sqrt{1 - v^2}}{1 + v\cos(\phi)}\,\nu_e$$

This differs in appearance from the Doppler shift equation given in Einstein's 1905 paper, but only because, in Einstein's equation, the angle ϕ is evaluated with respect to the emitter's rest frame, whereas in our equation the angle is evaluated with respect to the absorber's rest frame. These two angles differ because of the effect of aberration. If we let ϕ' denote the angle with respect to the emitter's rest frame, then ϕ' is related to ϕ by the aberration equation

$$\cos(\phi) = \frac{\cos(\phi') - v}{1 - v\cos(\phi)}$$

Substituting for cos(ϕ) into the previous equation gives Einstein's equation for the Doppler shift, i.e.,

$$\nu_a = \frac{1 - v\cos(\phi')}{\sqrt{1 - v^2}} \nu_e$$

Naturally for the "linear" cases, when ϕ = ϕ' = 0 or ϕ = ϕ' = π we have

$$\frac{\nu_a}{\nu_e} = \sqrt{\frac{1 - v}{1 + v}} \quad \text{or} \quad \frac{\nu_a}{\nu_e} \sqrt{\frac{1 + v}{1 - v}}$$

respectively. This highlights the symmetry between emitter and absorber that is so characteristic of relativistic physics.

Even more generally, consider an emitter moving with constant velocity u, an absorber moving with constant velocity v, and a signal propagating with velocity C in terms of an inertial coordinate system in which the signal's speed |C| is independent of direction. This would apply to a system of coordinates at rest with respect to the medium of the signal, and it would apply to any inertial coordinate system if the signal is light in a vacuum.

It would also apply to the case of a signal emitted at a fixed speed relative to the emitter, but only if we take u = 0, because in this case the speed of the signal is independent of direction only in terms of the rest frame of the emitter. We immediately have the relation

$$|r_a - r_e|^2 = |c|^2 (t_a - t_e)^2$$

where r_e and r_a are the position vectors of the emission and absorption events at the times t_e and t_a respectively. Differentiating both sides with respect to t_a and dividing through by $2(t_a - t_e)$, and noting that $(r_a - r_e)/(t_a - t_e) = C$, we get

$$C.\left(v - u\frac{dt_e}{dt_a}\right) = |C|^2 \left(1 - \frac{dt_e}{dt_a}\right)$$

where u and v are the velocity vectors of the emitter and absorber respectively. Solving for the ratio dt_e/dt_a, we arrive at the relation

$$\frac{dt_e}{dt_a} = \frac{|C|^2 - C.v}{|C|^2 - C.u} = \frac{1 - \dfrac{C.v}{|C|^2}}{1 - \dfrac{C.u}{|C|^2}}$$

Making use of the dot product identity r'"s = |r||s|cos ($\theta_{r,\,s}$) where $\theta_{r,\,s}$ is the angle between the r and s vectors, these can be re-written as

$$\frac{dt_e}{dt_a} = \frac{1 - \dfrac{|v|}{|C|}\cos(\theta_{C,v})}{1 - \dfrac{|u|}{|C|}\cos(\theta_{C,u})} = \frac{|C - v|\cos(\theta_{C,C-v})}{|C - u|\cos(\theta_{C,C-u})}$$

The frequency of any process is inversely proportional to the duration of the period, so the frequency at the absorber relative to the emitter, projected by means of the signal, is given by $\nu_a/\nu_e = dt_e/dt_a$. Therefore, the above expressions represent the classical Doppler effect for arbitrarily moving emitter and receiver. However, the elapsed proper time along a worldline moving with speed v in terms of any given inertial coordinate system differs from the elapsed coordinate time by the factor

$$\frac{d\tau}{dt} = \sqrt{1 - (v/c)^2}$$

where c is the speed of light in vacuum. Consequently, the actual ratio of proper times – and therefore proper frequencies – for the emitter and absorber is

$$\frac{\nu_a}{\nu_e} = \frac{d\tau_e}{d\tau_a} = \frac{1 - \dfrac{|v|}{|C|}\cos(\theta_{C,v})\sqrt{1 - (v/c)^2}}{1 - \dfrac{|u|}{|C|}\cos(\theta_{C,u})\sqrt{1 - (v/c)^2}}$$

The leading ratio is the classical Doppler effect, and the square root factor is the relativistic correction.

STELLAR ABERRATION

The aberration of starlight was discovered in 1727 by the astronomer James Bradley while he was searching for evidence of stellar parallax, which in principle ought to be observable if the Copernican theory of the solar system is correct.

He succeeded in detecting an annual variation in the apparent positions of stars, but the variation was not consistent with parallax. The observed displacement was greatest for stars in the direction perpendicular to the orbital plane of the Earth, and most puzzling was the fact that the displacement was exactly three months (i.e., 90 degrees) out of phase with the effect that would result from parallax due to the annual change in the Earth's position in orbit around the Sun.

It was as if he was expecting a sine function, but found instead a cosine function. Now, the cosine is the derivative of the sine, so this suggests that the effect he was seeing was not due to changes in the earth's *position*, but to changes in the Earth's (directional) *velocity*.

Indeed Bradley was able to interpret the observed shift in the incident angle of starlight relative to the Earth's frame of reference as being due to the transverse velocity of the Earth relative to the incoming corpuscles of light, assuming the latter to be moving with a finite speed c. The velocity of the corpuscles relative to the Earth equals their velocity vector c with respect to the Sun's frame of reference plus the negative of the orbital velocity vector v of the Earth, as shown below.

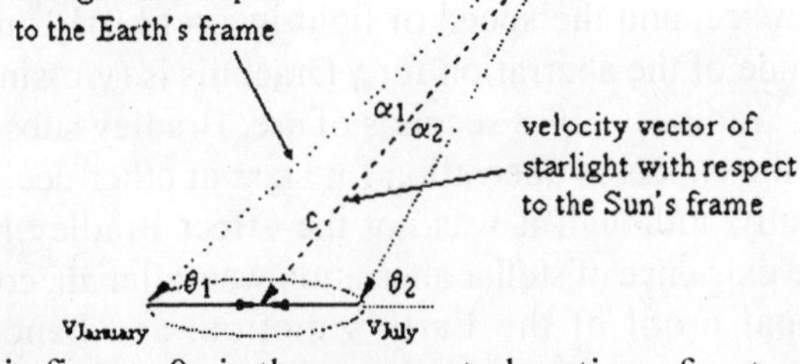

In this figure, θ_1 is the apparent elevation of a star above the Earth's orbital plane when the Earth's velocity is most directly toward the star (say, in January), and θ_2 is the apparent elevation six months later when the Earth's velocity is in the opposite direction. The law of sines gives

$$\sin(\alpha_1) = \sin(\theta_1)\frac{v_{January}}{c} \qquad \sin(\alpha_2) = \sin(\theta_2)\frac{v_{July}}{c}$$

Since the aberration angles α are quite small, we can closely approximate $\sin(\alpha)$ with just α. Therefore, the apparent position of a star that is roughly θ above the ecliptic ought to describe a small circle (or ellipse) around its true position, and the "radius" of this path should be $\sin(\theta)(v/c)$ where v is the Earth's orbital speed and c is the speed of light.

When Bradley made his discovery he was examining the star γ Draconis, which has a declination of about 51.5 degrees above the Earth's equatorial plane, and about 75 degrees above the ecliptic plane. Incidentally, most historical accounts say Bradley chose this star simply because it passes directly overhead in Greenwich England, the site of his observatory, which happens to be at about 51.5 degrees latitude.

Vertical observations minimize the effects of atmospheric refraction, but surely this is an incomplete explanation for choosing γ Draconis, because stars with this same declination range from 28 to 75 degrees above the ecliptic, due to the Earth's tilt of 23.5 degrees. Was it just a lucky coincidence that he chose (as Leibniz had previously) γ Draconis, a star with the maximum possible elevation above the ecliptic among stars that pass directly over Greenwich?

Accidental or not, he focused on nearly the ideal star for detecting aberration. The orbital speed of the Earth is roughly $v = (2.98)10^4$ m/sec, and the speed of light is $c = (3.0)10^8$ m/sec, so the magnitude of the aberration for γ Draconis is $(v/c)\sin(75 \text{ deg}) = (9.59)10^{-5}$ radians = 19.8 seconds of arc. Bradley subsequently confirmed the expected aberration for stars at other declinations.

Ironically, although it was not the effect Bradley had been seeking, the existence of stellar aberration was, after all, conclusive observational proof of the Earth's motion, and hence of the Copernican theory, which had been his underlying objective. Furthermore, the discovery of stellar aberration not only provided the first empirical proof of the Copernican theory, it also furnished a new and independent proof of the finite speed of light, and even enabled that speed to be estimated from knowledge of the orbital speed of the Earth. The result was consistent with the earlier estimate of the speed of light by Roemer based on observations of Jupiter's moons. Bradley's interpretation, based on the Newtonian corpuscular concept of light, accounted quite well for the basic phenomenon of stellar aberration.

However, if light consists of ballistic corpuscles their speeds ought to depend on the relative motion between the source and observer, and these differences in speed ought to be detectable, whereas no such differences were found.

For example, early in the 19th century Arago compared the focal length of light from a particular star at six-month intervals, when the Earth's motion should alternately add and subtract a velocity component equal to the Earth's orbital speed to the speed of light.

According to the corpuscle theory, this should result in a slightly different focal length through the system of lenses, but Arago observed no difference at all. In another experiment he viewed the aberration of starlight through a normal lens and through a thick prism with a very different index of refraction, which ought to give a slightly different aberration angle according to the Newtonian corpuscular model, but he found no difference.

Both these experiments suggest that the speed of light is independent of the motion of the source, so they tended to support the wave theory of light, rather than the corpuscular theory.

Unfortunately, the phenomenon of stellar aberration is somewhat problematic for theories that regard electromagnetic radiation as waves propagating in a luminiferous ether. It's worthwhile to examine the situation in some detail, because it is a nice illustration of the clash between mechanical and electromagnetic phenomena within the context of Galilean relativity.

If we conceive of the light emanating from a distant star reaching the Earth's location as a set of essentially parallel streams of particles normal to the Earth's orbit, then we have the situation shown in the left-hand figure below, and if we apply the Galilean transformation to a system of coordinates moving with the Earth (in the positive x direction) we get the situation shown in the right-hand figure.

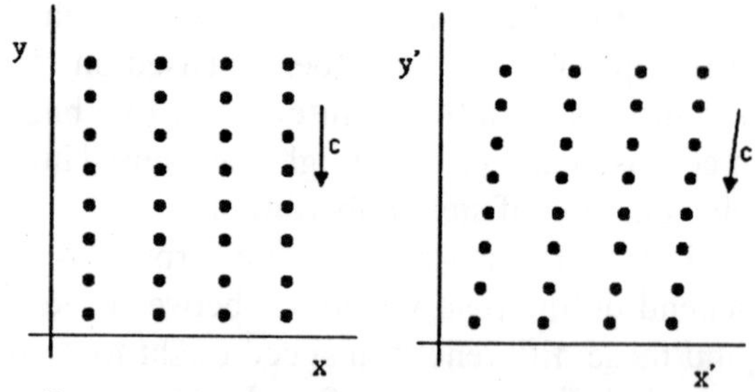

According to this model the aberration arises because each corpuscle has equations of motion of the form y = -ct and $x = x_0$, so the Galilean transformation x = x'+vt, y = y', t = t' leads to y' = −ct' and $x'+vt = x_0$, which gives (after eliminating t) the path x' − v(y'/c) = x_0. Thus we have dx'/dy' = v/c = tan(α). In contrast, if we conceive of the light as essentially a plane wave, the sequence of wave crests is as shown below.

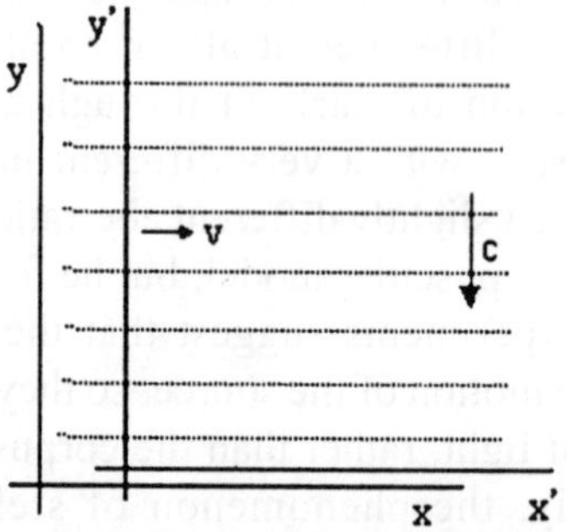

In this case each wavecrest has the equation y = −ct, with no x specification, because the wave is uniform over the entire wavefront. Applying the same Galilean transformation as before, we get simply y' = −ct', so the plane wave looks the same in terms of both systems of coordinates.

We might try to argue that the flow of energy follows definite streamlines, and if these streamlines are vertical with respect to the unprimed coordinates they would transform into slanted streamlines in the primed coordinates, but this would imply that the direction of propagation of the wave energy is not exactly normal to the wave fronts, in conflict with Maxwell's equations.

This highlights the incompatibility between Maxwell's equations and Galilean relativity, because if we regard the primed coordinates as stationary and the distant star as moving transversely with speed –v, then the waves reaching the Earth at this moment should have the same form as if they were emitted from the star when it was to the right of its current position, and therefore the wave fronts ought to be slanted by an angle of v/c.

Of course, we do actually observe aberration of this amount, so the wave fronts really must be tilted with respect to the primed coordinates, and we can fairly easily explain this in terms of the wave model, but the explanation leads to a new complication.

According to the early 19th century wave model with a stationary ether, an observation of a distant star consists of focusing a set of parallel rays from that star down to a point, and this necessarily involves some propagation of light in the transverse direction (in order to bring the incoming rays together).

Taking the focal point to be midway between two rays, and assuming the light propagates transversely at the same speed in both directions, we will align our optical device normal to the plane wave fronts. However, suppose the effective speed of light is slightly different in the two transverse directions.

If that were the case, we would need to tilt our optical device, and this would introduce a time skew in our evaluation of the wave front, because our optical image would associate rays from different points on the wave front at slightly different times. As a result, what we regard as the wave front would actually be slanted.

The proponents of the wave model argued that the speed of light is indeed different in the two transverse directions relative to a telescope on the Earth pointed up at a star, because the Earth is moving sideways (through the ether) with respect to the incoming rays.

Assuming light always propagates at the fixed speed c relative to the ether, and assuming the Earth is moving at a speed v relative to the ether, we could argue that the transverse speed of light inside our telescope is c+v in one direction and c–v in the other.

To assess the effect of this asymmetry, consider for simplicity just two mirror elements of a reflecting telescope, focusing incoming rays as illustrated below.

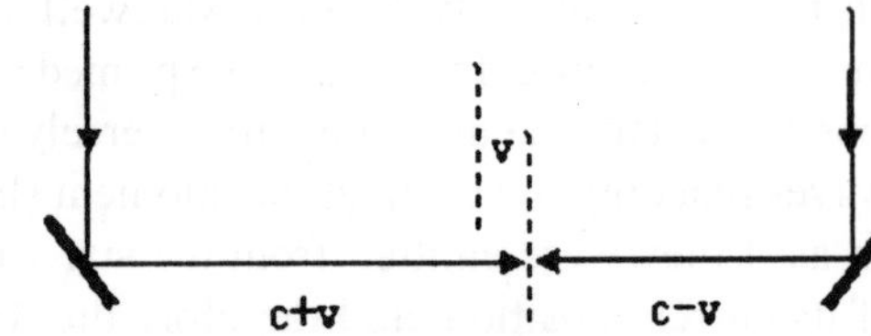

The two incoming rays shown in this figure are from the same wavecrest, but they are not brought into focus at the midpoint of the telescope, due to the (putative) fact that the telescope is moving sideways through the ether with a speed v. Both pulses strike the mirrors at the same time, but the left hand pulse goes a distance proportional to c+v in the time it takes the right hand pulse to go a distance proportional to c–v.

In order to bring the wave crest into focus, we need to increase the path length of the left hand ray by a distance proportional to v, and decrease the right hand path length by the same distance. This is done by tilting the telescope through a small angle whose tangent is roughly v/c, as shown below.

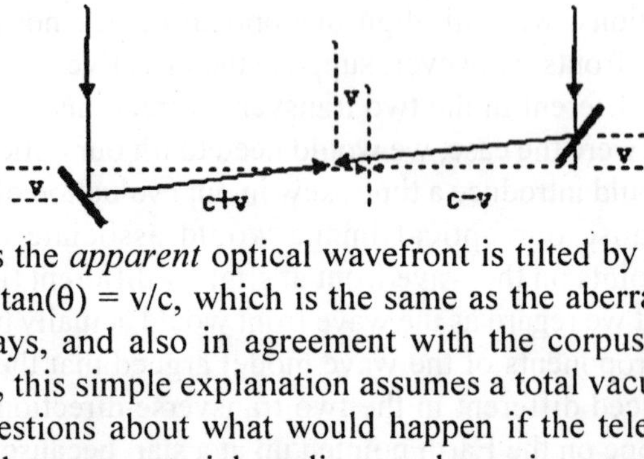

Thus the *apparent* optical wavefront is tilted by an angle θ given by tan(θ) = v/c, which is the same as the aberration angle for the rays, and also in agreement with the corpuscle model. However, this simple explanation assumes a total vacuum, and it raises questions about what would happen if the telescope was filled with some material medium such as air or water. It was already accepted in Fresnel's day, for both the wave and the corpuscle models of light, that light propagates more slowly in a dense medium than in vacuum. Specifically, the speed of light in a medium with index of refraction n is c/n.

Hence if we fill our reflecting telescope with such a medium, then the speed of light in the two transverse directions would be c/n + v and c/n – v, and the above analysis would lead us to expect an aberration angle given by tan(θ) = nv/c. The index of refraction of air is just 1.0003, so this doesn't significantly affect the observed aberration angle for telescopes in air. However, the index of refraction of water is 1.33, so if we fill a telescope with water, we ought to observe (according to this theory) significantly more stellar aberration. Such experiments have actually been carried out, but no effect on the aberration angle is observed.

In 1818 Fresnel suggested a way around this problem. His hypothesis, which he admitted appeared extraordinary at first sight, was that although the luminiferous ether through which light propagates is nearly immobile, it is dragged along slightly by material objects, and the higher the refractive index of the object, the more it drags the ether along with its motion. If an object with refractive index n moves with speed v relative to the nominal rest frame of the ether, Fresnel hypothesized that the ether inside the object is dragged forward at a speed $(1 - 1/n^2)v$.

Thus for objects with n = 1 there is no dragging at all, but for n greater than 1 the ether is pulled along slightly. Fresnel gave a plausibility argument based on the relation between density and refractivity, making his hypothesis seem at least slightly less contrived, although it was soon pointed out that since the index of refraction of a given medium varies with frequency, Fresnel's model evidently *requires a different ether for each frequency.* Neglecting this second-order effect of chromatic dispersion, Fresnel was able on the basis of his partial dragging hypothesis to account for the absence of any change in stellar aberration for different media. He pointed out that, in the above analysis, the speed of light in the two directions has the values

$$\left[\frac{c}{n} - v\left(1 - \frac{1}{n^2}\right)\right] + v = \frac{c}{n} + \frac{v}{n^2} \qquad \left[\frac{c}{n} + v\left(1 - \frac{1}{n^2}\right)\right] - v = \frac{c}{n} + \frac{v}{n^2}$$

For the vacuum we have n = 1, and these expressions are the same as before. In the presence of a material medium with n greater than 1, the optical device must now be tilted through an angle whose tangent is approximately

$$\frac{v/n^2}{c/n} = \frac{1}{n}\left(\frac{v}{c}\right)$$

It might seem as if Fresnel's hypothesis has simply resulted in exchanging one problem for another, but recall that our telescope is aligned normal to the apparent wave front, whereas it is at an angle of v/c to the normal of the actual wave front, so the wave will be refracted slightly. According to Snell's law (which for small angles is $n_1\theta_1 = n_2\theta_2$), the refracted angle will be less than the incident angle by the factor 1/n. Hence we must orient our telescope at an angle of v/c in order for the rays within the medium to be at the required angle.

This is how, on the basis of somewhat adventuresome hypotheses and assumptions, physicists of the 19th century were able to account for stellar aberration on the basis of the wave model of light. (Accommodating the lack of effect of differing indices of refraction proved to be even more challenging for the corpuscular model). Fresnel's remarkable hypothesis was directly confirmed by Fizeau, and it is now recognized as a first-order approximation of the relativistic velocity addition law, composing the speed of light in a medium with the speed of the medium

$$\frac{\frac{c}{n}+u}{1+\frac{1}{c^2}\left(\frac{c}{n}u\right)}$$

$$\approx\left(\frac{c}{n}+u\right)\left(1-\frac{1}{n}\frac{c}{n}\right)=\frac{c}{n}+u-\frac{u}{n^2}-\frac{u^2}{cn}\approx\frac{c}{n}+u\left(1-\frac{1}{n^2}\right)$$

It's worth noting that all the "speeds" discussed here are phase speeds, corresponding to the time parameter for a given wave. Lorentz later showed that Fresnel's formula could also be interpreted in the context of a perfectly immobile ether along with the assumption of phase shifts in the incoming wave fronts so that the *effective* time parameter transformation was not the Galilean $t' = t$ but rather $t' = t - vx/c^2$.

Despite the success of Fresnel's hypothesis in matching all optical observations to the first order in v/c, many physicists considered his partially dragged ether model to be ad hoc and unphysical (especially the apparent need for a different ether for each frequency of light), so they sought other explanations for stellar aberration that would be consistent with a more mechanistically realistic wave model.

As an alternative to Fresnel's hypothesis, Lorentz evaluated a proposal of Stokes, who in 1846 had suggested that the ether is *totally* dragged along by material bodies (so the ether is co-moving with the body at the body's surface), and is irrotational, incompressible, and inviscid, so that it supports a velocity potential.

Under these assumptions it can be shown that the normal of a light wave incident on the Earth undergoes a total deflection during its approach such that (to first order) the apparent shift in the star's position agrees with observation. Unfortunately, as Lorentz pointed out, the assumptions of Stokes' theory are mutually contradictory, because the potential flow field around a sphere does not give zero velocity on the sphere's surface. Instead, the velocity of the ether wind on the Earth's surface would vary with position, and so too would the aberration of starlight. Planck suggested a way around this objection by supposing the luminiferous ether was compressible, and accumulated with greatly increased density around large objects.

Lorentz admitted that this was conceivable, but only if we also assume the speed of light propagating through the ether is unaffected by the changes in density of the ether, an assumption that plainly contradicts the behaviour of wave propagation in ordinary substances. He concluded

In this branch of physics, in which we can make no progress without some hypothesis that looks somewhat startling at first sight, we must be careful not to rashly reject a new idea... The hypothesis that the velocity of light is not in the least altered by it, is not very satisfactory.

With the failure of Stoke's theory, the only known way of reconciling stellar aberration with a wave theory of light was Fresnel's "extraordinary" hypothesis of partial dragging, or Lorentz's equivalent interpretation in terms of the effective phase time parameter t'. However, the Fresnel-Lorentz theory predicted a non-null result for the Michelson-Morley experiment, which was the first experiment accurate to the second order in v/c. To remedy this, Lorentz ultimately incorporated Fitzgerald's length contraction into his theory, which amounts to replacing the Galilean transformation $x' = x - vt$ with the relation $x' = (x - vt)/$

$(1 - (v/c)^2)^{1/2}$, and then for consistency applying this same second-order correction to the time transformation, giving $t' = (t - vx/c^2)/(1 - (v/c)^2)^{1/2}$, thereby arriving at the full Lorentz transformation. By this point the posited luminiferous ether had lost all of its mechanistic properties.

Meanwhile, Einstein's 1905 paper on the electrodynamics of moving bodies included a greatly simplified derivation of the full Lorentz transformation, dispensing with the ether altogether, and analyzing a variety of phenomena, including stellar aberration, from a purely kinematical point of view. If a photon is emitted from object A at the origin of the xyt coordinates and an angle α relative to the x axis, then at time t_1 it will have reached the point

$$x_1 = t_1 \cos(\alpha) \quad y_1 = t_1 \sin(\alpha) \quad t_1 = t_1$$

(Notice that the units have been scaled to make c = 1, so the Minkowski metric for a null interval gives $x_1^2 + y_1^2 = t_1^2$). Now consider an object B moving in the positive x direction with velocity v, and being struck by the photon at time t_1 as shown below.

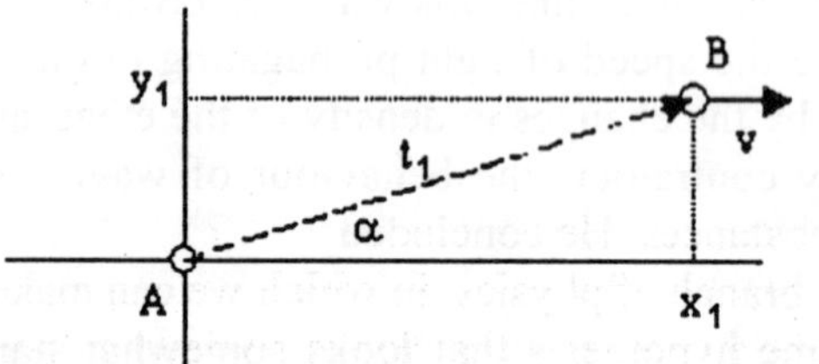

Naturally an observer riding along with B will not see the light ray arriving at an angle α from the x axis, because according to the system of coordinates co-moving with B the source object A has moved in the x direction (but not in the y direction) between the times of transmission and reception of the photon.

Since the angle is just the arctangent of the ratio of Δy to Δx of the photon's path, and since value of Δx is different with respect to B's co-moving inertial coordinates whereas Δy is the same, it's clear that the angle of the photon's path is different with respect to B's co-moving coordinates than with respect to A's co-moving coordinates. In general the transformation of the angles of the paths of moving objects from one system of inertial coordinates to another is called aberration. To determine the angle of the incoming ray with respect to the co-moving inertial coordinates of B, let x'y't' be an orthogonal coordinate system aligned with the xyt coordinates but moving in the positive x direction with velocity v, so that B is at rest in the primed coordinate system.

Without loss of generality we can co-locate the origins of the primed and unprimed coordinates systems, so in both systems the photon is emitted at (0, 0, 0). The endpoint of the photon's path in the primed coordinates can be computed from the unprimed coordinates using the standard Lorentz transformation for a boost in the positive x direction:

$$x_1' = \frac{x_1 - vt}{\sqrt{1-v^2}} = t_1 \frac{\cos(\alpha) - v}{\sqrt{1-v^2}}$$

$$y_1' = y_1 = t_1 \sin(\alpha)$$

$$t_1' = \frac{t_1 - vx_1}{\sqrt{1-v^2}} = t_1 \frac{1 - v\cos(\alpha) - v}{\sqrt{1-v^2}}$$

Just as we have $\cos(\alpha) = x_1/t_1$, we also have $\cos(\alpha') = x_1'/t_1'$, and so

$$\cos(\alpha') = \frac{\cos(\alpha) - v}{1 - v\cos(\alpha)}$$

which is the general relativistic aberration formula relating the angles of light rays with respect to relatively moving coordinate systems. Likewise we have $\sin(\alpha') = y_1'/t_1'$, from which we get

$$\sin(\alpha') = \frac{\sin(\alpha)\sqrt{1-v^2}}{1 - v\cos(\alpha)}$$

Using these expressions for the sine and cosine of α' it follows that

$$\frac{1+\cos(\alpha')}{\sin(\alpha')} = \frac{1+\cos(\alpha)}{\sin(\alpha)}\sqrt{\frac{1-v}{1+v}}$$

Recalling the trigonometric identity tan(z) = sin(2z)/[1+cos(2z)] this gives

$$\tan(\alpha'/2) = \sqrt{\frac{1+v}{1-v}}\tan(\alpha'/2)$$

which immediately shows that aberration can be represented by stereographic projection from a sphere to the tangent plane.

To see the effect of equation, suppose that, with respect to the inertial rest frame of a given particle, the rays of starlight incident on the particle are uniformly distributed in all directions. Then suppose the particle is given some speed v in the positive x direction relative to this original isotropic frame, and we evaluate the angles of incidence of those same rays of starlight with respect to the particle's new rest frame. The results, for speeds ranging from 0 to 0.999, are shown in the figure below.

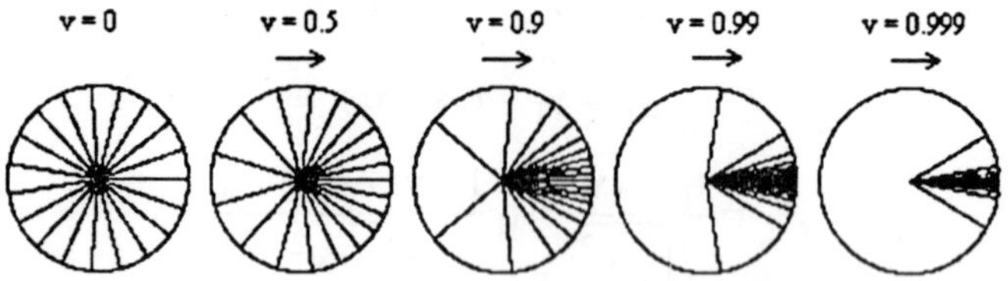

The preceding derivation applies to the case when the light is emitted from the unprimed coordinate system at a certain angle and evaluated with respect to the primed coordinate system, which is moving relative to the unprimed system. If instead the light was emitted from B and received at A, we can repeat the above derivation, except that the direction of the light ray is reversed, going now from B to A.

The spatial coordinates are all the same but the emission event now occurs at $-t_1$, because it is in the past of event (0, 0, 0). The result is simply to replace each occurrence of v in the above expressions with -v. Of course, we could reach the same result simply by transposing the primed and unprimed angles in the above expressions. This configuration corresponds to a distant star at A sending starlight to the Earth at B, which is moving nearly perpendicular to the incoming ray.

Incidentally, the aberration formula used by astronomers to evaluate the shift in the apparent positions of stars resulting from the Earth's orbital motion is often expressed in terms of angles with respect to the y axis (instead of the x axis), as shown below:

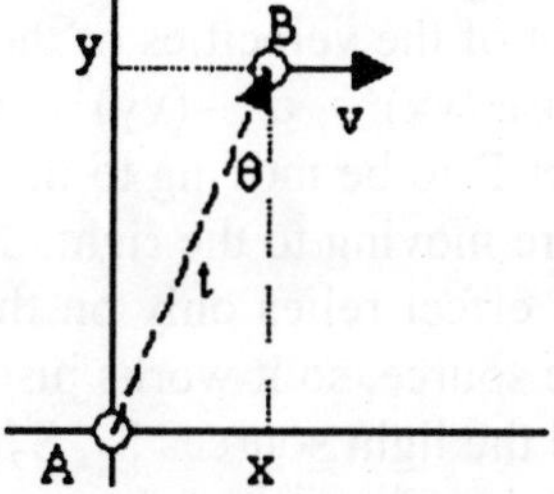

This gives the greatest aberration effect, which explains why the stars furthest from the ecliptic plane experience the greatest aberration. The formula can be found simply by making the substitution $\alpha = \pi - \theta$ in equation, and noting the trigonometric identity $\tan(\mathrm{acos}(\pi/2 - x)) = x/\sqrt{1-x^2}$. This gives the equivalent form

$$\tan(\alpha) = \frac{\cos(\alpha') + v}{\sqrt{1-x^2}\sqrt{1-x^2(\alpha')^2}}$$

Another interesting aspect of aberration is illustrated by considering two separate light sources S_1 and S_2, and two momentarily coincident observers A and B as shown below:

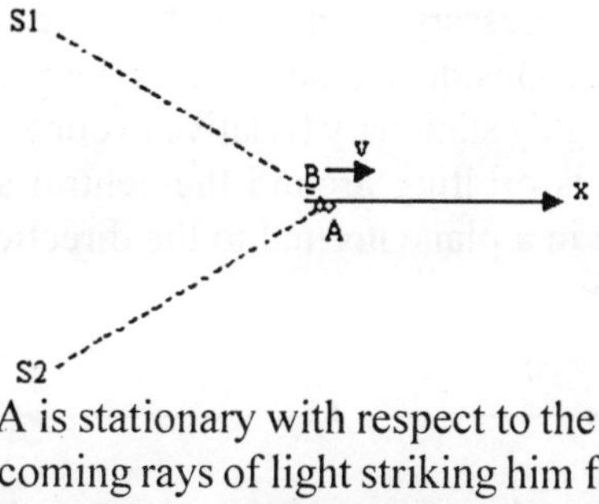

If observer A is stationary with respect to the sources of light, he will see the incoming rays of light striking him from the negative x direction. Thus, the light will impart a small amount of momentum to observer A in the positive x direction. On the other hand, suppose observer B is moving to the right (away from the sources of light) at nearly the speed of light.

According to our aberration formula, if B is traveling with a sufficiently great speed, he will see the light from S_1 and S_2 approaching from the *positive* x direction, which means that the photons are imparting momentum to B in the *negative* x direction - even though the light sources are "behind" B. This may seem paradoxical, but the explanation becomes clear when we realise that the x component of the velocities of the incoming light rays is less than c (because $(vx)^2 = c^2 - (vy)^2$), which means that it's possible for observer B to be moving to the right *faster* than the incoming photons are moving to the right.

Of course, this effect relies only on the relative motion of the observer and the source, so it works just as well if we regard B as motionless and the light sources S_1, S_2 moving to the left at near the speed of light. Thus, it might seem that we could use light rays to "pull" an object from behind, and in a sense this is true. However, since the light rays are moving to the right more slowly than the object, they clearly cannot catch up with the object from behind, so they must have been emitted when the object was still to the left of the sources.

This illustrates how careful one must be to correctly account for the effective aberration of non-uniformly moving objects, because the simple aberration formulas are based on the assumption that the light source has been in uniform motion for an indefinite period of time.

To correctly describe the aberration of non-uniformly moving light sources it is necessary to return to the basic metrical relations.

For example, consider a binary star system in which one large central star is roughly stationary (relative to our Sun), and a smaller companion star is orbiting around the central star with a large angular velocity in a plane normal to the direction to our Sun, as illustrated below.

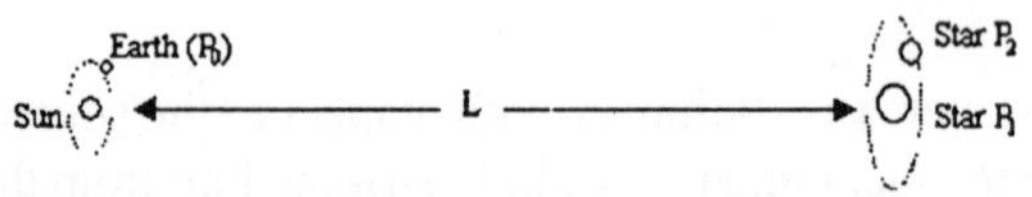

It might seem that the periodic variations in the velocity of the smaller star relative to our Sun would result in significantly different amounts of aberration as viewed from the Earth, causing the two components of the binary star system to appear in separate locations in the sky - which of course is not what is observed. Fortunately, it's easy to show that the correct application of the principles of special relativity, accounting for the non-uniform variations in the orbiting star's velocity, leads to prediction that agree perfectly with observation of binary star systems. At any moment of observation on Earth we can consider ourselves to be at rest at the point P_0 in the momentarily co-moving inertial frame, with respect to which our coordinates are

$$x_0(t) = 0 \quad y_0(t) = 0 \quad z_0(t) = 0 \quad t_0(t) = t$$

Suppose the large central star of a binary pair is at point P_1 at a distance L from the Earth with the coordinates

$$x_1(t) = -vt \quad y_1(t) = 0 \quad z_1(t) = L \quad t_1(t) = t$$

The fundamental assertion of special relativity is that light travels along null paths, so if a pulse of light is emitted from the star at time t = T and arrives at Earth at time t = 0, we have

$$T^2 = (vT)^2 + L^2$$

and so

$$T = \frac{L}{\sqrt{1-v^2}}$$

from which it follows that x_1/z_1 at time T is $\sqrt{1-v^2}$. Thus, for the central star we have the aberration angle

$$\alpha = \text{invtan}\left(\frac{v}{\sqrt{1-v^2}}\right) = \text{inv}\sin(v)$$

Now, what about the aberration of the other star in the binary pair, the one that is assumed to be much smaller and revolving at a radius R and angular speed w around the larger star in a plane perpendicular to the Earth? The coordinates of that revolving star at point P_2 are

$$x_2(t) = -vt + R\cos(\theta) \quad y_2(t) = R\sin(\theta) \quad z_2(t) = L \quad t_2(t) = t$$

where θ = wt is the angular position of the smaller star in its orbit. Again, since light travels along null paths, a pulse of light arriving on Earth at time t = 0 was emitted at time t = T satisfying the relation

$$\begin{aligned} T^2 &= [-vT + R\cos(\theta)]^2 + [R\sin(\theta)]^2 + L^2 \\ &= v^2T^2 - 2vTR\cos(\theta) + R^2 + L^2 \end{aligned}$$

Solving this quadratic for T (and noting that the phase θ depends entirely on the arbitrary initial conditions of the orbit) gives

$$T = \frac{Rv\cos(\theta)}{1-v^2} + \frac{L}{\sqrt{1-v^2}}\sqrt{1+\left(\frac{R}{L}\right)\left(1+\frac{v^2}{1-v^2}\cos(\theta)\right)}$$

If the radius R of the binary star's orbit is extremely small in comparison with the distance L from those stars to the Earth, and assuming v is not very close to the speed of light, then the quantity inside the square root is essentially equal to 1. Therefore, the tangents of the angles of incidence in the x and y directions are

$$\frac{x_2}{z_2} = \frac{R}{L}\cos(\theta) + \frac{v}{\sqrt{1-v^2}} - \frac{R}{L}\frac{v^2}{1-v^2}\cos(\theta)$$

$$\frac{y_2}{z_2} = \frac{R}{L}\sin(\theta)$$

These expressions make it clear why Einstein emphasized in his 1905 treatment of aberration that the light source was at infinite distance, i.e., L goes to infinity, so all but the middle term of the x tangent vanish. Of course, the leading terms in these tangents are obviously just the inherent "static" angular separation between the two stars viewed from the Earth, and the last term in the x tangent is completely negligible assuming R/L and/or v are sufficiently small compared with 1, so the aberration angle is essentially

$$\alpha = \text{inv}\tan\left(\frac{v}{\sqrt{1-v^2}}\right) = \text{inv}\sin(v)$$

which of course is the same as the aberration of the central star. Indeed, binary stars have been carefully studied for over a century, and the aberrations of the components are consistent with the relativistic predictions for reasonable Keplerian orbits. (Incidentally, recall that Bradley's original formula for aberration was $\tan(\alpha) = v$, whereas the corresponding relativistic equation is $\sin(\alpha) = v$. The actual aberration angles for stars seen from Earth are small enough that the sine and tangent are virtually indistinguishable). The experimental results of Michelson and Morley, based on beams of light pointed in various directions with respect to the Earth's motion around the Sun, can also be treated as aberration effects. Let the arm of Michelson's interferometer be of length L, and let it make an angle α with the direction of motion in the rest frame of the arm.

We can establish inertial coordinates t, x, y in this frame, in terms of which the light pulse is emitted at $t_1 = 0$, $x_1 = 0$, $y_1 = 0$, reflected at $t_2 = L$, $x_2 = L\cos(\alpha)$, $y_2 = L\sin(\alpha)$, and arrives back at the origin at $t_3 = 2L$, $x_3 = 0$, $y_3 = 0$.

The Lorentz transformation to a system x', y', t' moving with velocity v in the x direction is $x' = (x-vt)/\gamma$, $y' = y$, $t' = (t-vx)/\gamma$ where $\gamma^2 = (1-v^2)$, so the coordinates of the three events are $x_1' = 0$, $y_1' = 0$, $t_1' = 0$, and $x_2' = L(\cos(\alpha)-v)/\gamma$, $y_2' = L\sin(\alpha)$, $t_2' = L[1-v\cos(\alpha)]/\gamma$, and $x_3' = -2vL/\gamma$, $y_3' = 0$, $t_3' = 2L/\gamma$. Hence the total elapsed time in the primed coordinates is $2L/\gamma$. Also, the total spatial distance traveled is the sum of the outward distance

$$\sqrt{(x_2' - x_1)^2 + (x_2' - x_1)^2} = \frac{L - 2v\cos(\alpha)}{\sqrt{1-v^2}}$$

and the return distance

$$\sqrt{(x_3' - x_2')^2 + (y_3' - y_2')^2} = \frac{L - 2v\cos(\alpha)}{\sqrt{1-v^2}}$$

so the total distance is $2L/\gamma$, giving a light speed of 1 regardless of the values of v and α. Of course, the angle of the interferometer arm cannot be α with respect to the primed coordinates.

The tangent of the angle equals the arm's y extent divided by its x extent, which gives $\tan(\alpha) = L\sin(\alpha)/[L(\cos(\alpha)]$ in the arm's rest coordinates. In the primed coordinates the y' extent of the arm is the same as the y extent, $L\sin(\alpha)$, but the x' extent is $L\cos(\alpha)\gamma$, so the tangent of the arm's angle is $\tan(\alpha') = \tan(\alpha)/\gamma$.

However, this should not be confused with the angle (in the primed coordinates) of the light pulse as it travels along the arm, because the arm is in motion with respect to the primed coordinates. The outward direction of motion of the light pulse is given by evaluating the primed coordinates of the emission and absorption events at x_1, y_1 and x_2, y_2 respectively. Likewise the inward direction of the light pulse is based on the interval from x_2, y_2 to x_3, y_3. These give the tangents of the outward and inward angles

$$\tan\left(\beta_{out}'\right) = \frac{\sin(\alpha)\sqrt{1-v^2}}{\cos(\alpha) - v} \qquad \tan\left(\beta_{in}'\right) = \frac{\sin(\alpha)\sqrt{1-v^2}}{\cos(\alpha) + v}$$

Naturally these are consistent with the result of taking the ratio of equations.

MOBIUS TRANSFORMATIONS AND THE NIGHT SKY

Any proper orthochronous Lorentz transformation (including ordinary rotations and relativistic boosts) can be represented by

$$X \to QXQ^*$$

where

$$x = \begin{bmatrix} t+z & t+iy \\ x-iy & t-z \end{bmatrix} \qquad Q = \begin{bmatrix} a & b \\ c & d \end{bmatrix}$$

and Q* is the transposed conjugate of Q. The coefficients a, b, c, d of Q are allowed to be complex numbers, normalized so that ad – bc = 1. Just to be explicit, this implies that if we define

$$a = a_1 + ia_2 \quad b = b_1 + ib_2 \quad c = c_1 + ic_2 \quad d = d_1 + id_2$$

then the Lorentz transformation is

$$t' = [a_1b_1 + a_2b_2 + c_1d_1 + c_2d_2]x + [a_1b_2 - a_2b_1 + c_1d_2 - c_2d_1]y$$

$$+(1/2)\left[|a|^2 - |b|^2 + |c|^2 - |d|^2\right]z + (1/2)\left[|a|^2 + |b|^2 + |c|^2 + |d|^2\right]t$$

$$x' = [a_1d_1 + a_2d_2 + b_1c_1 + b_2c_2]x + [a_1d_2 - a_2d_1 - b_1c_2 + b_2c_1]y$$

$$+[a_1c_1 + a_2c_2 - b_1d_1 - b_2d_2]z + [a_1c_1 + a_2c_2 + b_1d_1 + b_2d_2]t$$

$$y' = [a_1d_2 + a_2d_1 - b_1c_2 + b_1c_2]x + [a_1d_1 - a_2d_2 - b_1c_1 - b_2c_2]y$$

$$+[a_1c_2 + a_2c_1 - b_1d_2 - b_2d_1]z + [-a_1c_2 + a_2c_1 - b_1d_2 + b_2d_1]t$$

$$z' = [a_1b_1 + a_2b_2 - c_1d_1 - c_2d_2]x + [a_1b_2 - a_2b_1 - c_1d_1 + c_2d_1]y$$

$$+(1/2)\left[|a|^2 - |b|^2 - |c|^2 + |d|^2\right]$$

$$z + (1/2)\left[|a|^2 + |b|^2 - |c|^2 - |d|^2\right]t$$

Two observers at the same point in spacetime but with different orientations and velocities will "see" incoming light rays arriving from different relative directions with respect to their own frames of reference, due partly to ordinary rotation, and partly to the aberration effect. This leads to the remarkable fact that the combined effect of any proper orthochronous (and homogeneous) Lorentz transformation on the incidence angles of light rays at a point corresponds precisely to the effect of a particular linear fractional transformation on the Riemann sphere via ordinary stereographic projection from the extended complex plane. The latter is illustrated below:

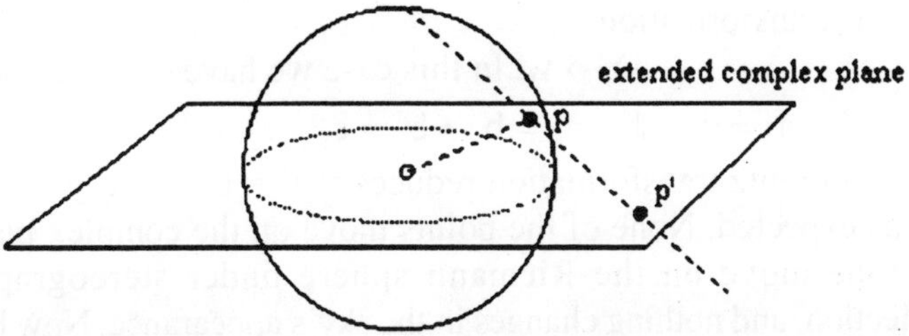

The complex number p in the extended complex plane is identified with the point p' on the unit sphere that is struck by a line from the "North Pole" through p. In this way we can identify each complex number uniquely with a point on the sphere, and vice versa. (The North Pole is identified with the "point at infinity" of the extended complex plane, for completeness).

Relative to an observer located at the centre of the Riemann sphere, each point of the sphere lies in a certain direction, and these directions can be identified with the directions of incoming light rays at a point in spacetime.

If we apply a Lorentz transformation of the form this observer, specified by the four complex coefficients a, b, c, d, the resulting change in the directions of the incoming rays of light is given exactly by applying the linear fractional transformation (also known as a Mobius transformation)

$$w \to \frac{aw + b}{cw + d} \qquad a,b,c,d \in C, \quad ad - bc = 1$$

to the points of the extended complex plane. Of course, our normalization ad – bc = 1 implies the two conditions

$$a_1d_1 - a_2d_2 - b_1c_1 + b_2c_2 = 1$$

$$a_1 d_2 - a_2 d_1 - b_1 c_2 + b_2 c_1 = 0$$

so of the eight coefficients needed to specify the four complex numbers a, b, c, d, these two constraints reduce the degrees of freedom to six, which is precisely the number of degrees of freedom of Lorentz transformations (namely, three velocity components v_x, v_y, v_z, and three angular specifications for the longitude and latitude of our line of sight and orientation about that line).

To illustrate this correspondence, first consider the "identity" Mobius transformation

w → w. In this case we have

$$a_1 = d_1 = 1 \quad a_2 = b_1 = b_2 = c_1 = c_2 = d_2 = 0$$

so our Lorentz transformation reduces to t' = t, x' = x, y' = y, z' = z as expected. None of the points move on the complex plane, so none move on the Riemann sphere under stereographic projection, and nothing changes in the sky's appearance. Now let's consider the Mobius transformation w → −1/w. In this case we have

$$b_1 = -1 \quad c_1 = 1 \qquad a_1 = a_2 = b_2 = c_2 = d_1 = d_2 = 0$$

and so the corresponding Lorentz transformation is t' = t, x' = −x, y' = y, z' = −z . Thus the x and z coordinates have been reflected. This is certainly a proper orthochronous Lorentz transformation, because the determinant is +1 and the coefficient of t is positive. But does reflecting the x and z coordinates agree with the stereographic effect on the Riemann sphere of the transformation w → −1/w? Note that the point w = r + 0i maps to −1/r + 0i.

There's a nice little geometric demonstration that the stereographic projections of these points have coordinates (x, 0, z) and (−x, 0, −z) respectively, noting that the two projection lines have negative inverse slopes and so are perpendicular in the xz plane, which implies that they must strike the sphere on a common diameter (by Pythagoras' theorem).

A similar analysis shows that points off the real axis with projected coordinates (x, y, z) in general map to points with projections (−x, y, −z) points. The two examples just covered were both trivial in the sense that they left t unchanged. For a more interesting example, consider the Mobius transformation $w \to w + p$, which corresponds to the Lorentz transformation

$$t' = \left(1 + p^2/2\right)t + px - \left(p^2/2\right)z$$

$$x' = pt + x - pz$$

$$y' = y$$

$$z' = \left(p^2/2\right)t + px + \left(1 - p^2/2\right)z$$

If we denote our spacetime coordinates by the column vector X with components $x_0 = t$, $x_1 = x$, $x_2 = y$, $x_3 = z$, then the transformation can be written as

$$X' = LX$$

where

$$L = \begin{bmatrix} 1 + p^2/2 & p & 0 & -p^2/2 \\ p & 1 & 0 & -p \\ 0 & 0 & 1 & 0 \\ p^2/2 & p & 0 & 1 - p^2/2 \end{bmatrix}$$

To analyse this transformation it's worthwhile to note that we can decompose any Lorentz transformation into the product of a simple boost and a simple rotation. For a given relative velocity with magnitude $|v|$ and components v_1, v_2, v_3, let γ denote the "boost factor"

$$\gamma = \frac{1}{\sqrt{1 - |v|^2}}$$

It's clear that

$$L_{00} = \gamma \quad L_{10} = \gamma v_1 \quad L_{20} = \gamma v_2 \quad L_{30} = \gamma v_3$$

Thus, these four components of L are fixed purely by the boost. The remaining components depend on the rotational part of the transformation.

If we define a "pure boost" as a Lorentz transformation such that the two frames see each other moving with velocities (v_1, v_2, v_3) and $(-v_1, -v_2, -v_3)$ respectively, then there is a unique pure boost for any given relative velocity vector v_1, v_2, v_3. This boost has the components

$$B\begin{bmatrix} \gamma & \gamma v_1 & \gamma v_2 & \gamma v_3 \\ Qv_1v_0 & 1+Qv_1^{\ 2} & Qv_1v_2 & Qv_1v_3 \\ Qv_2v_0 & Qv_2v_0 & 1+Qv_2^{\ 2} & Qv_2v_3 \\ Qv_3v_0 & Qv_3v_1 & Qv_3v_2 & 1+Qv_2^{\ 2} \end{bmatrix}$$

where $Q = (\gamma-1)/|v|^2$. From our expression for L we can identify the components to give the boost velocity in terms of the Mobius parameter p

$$v_1 = \frac{2p}{2+p^2} \quad v_2 = 0 \quad v_3 = \frac{p^2}{2+p^2}$$

and

$$|v|^2 = \left(\frac{p^2}{2+p^2}\right)(4+p^2) \quad \gamma = 1 + \frac{p^2}{2} \quad Q = \frac{(2+p^2)^2}{2(4+p^2)}$$

From these we write the pure boost part of L as follows

$$B = \begin{bmatrix} 1+\dfrac{p^2}{2} & p & 0 & \dfrac{p^2}{2} \\ p & 1+\dfrac{2p^2}{4+p^2} & 0 & \dfrac{p^3}{4+p^3} \\ 0 & 0 & 1 & 0 \\ p^2/2 & \dfrac{p^3}{4+p^2} & 0 & 1+\dfrac{p^4}{2(4+p^2)} \end{bmatrix}$$

We know that our Lorentz transformation L can be written as the product of this pure boost B times a pure rotation R, i.e., L = BR, so we can determine the rotation

$$R = B^{-1}L$$

which in this case gives

$$R = \begin{bmatrix} 1 & 0 & 0 & 0 \\ 0 & \dfrac{4-p^3}{4+p^2} & 0 & \dfrac{-4p}{4+p^2} \\ 0 & 0 & 1 & 0 \\ 0 & \dfrac{4p}{4+p^2} & 0 & \dfrac{4-p^2}{4+p^2} \end{bmatrix}$$

In terms of Euler angles, this represents a rotation about the y axis through an angle of

$$\theta = \text{inv}\cos\left(\frac{4-[^2}{4+p^2}\right)$$

The correspondence between the coefficients of the Mobius transformation and the Lorentz transformation described above assumes stereographic projection from the North pole to the equatorial plane. More generally, if we're projecting from the North Pole of the Riemann sphere to a complex plane parallel to (but not necessarily on) the equator, and if the North Pole is at a height h above the plane, then every point in the plane is a factor of h further away from the origin than in the case of equatorial projection (h=1), so the Mobius transformation corresponding to the above Lorentz transformation is w → (Aw+B)/(Cw+D) where

$$A = a\sqrt{h} \qquad B=\frac{b}{\sqrt{h}} \qquad C=c\sqrt{h} \qquad D=\frac{d}{\sqrt{h}}$$

It's also worth noting that the instantaneous aberration observed by an *accelerating* observer does not differ from that observed by a momentarily co-moving inertial observer. We're referring here to the null (light-like) rays incident on a *point* of zero extent, so this is not like a finite spinning body whose outer edges have significant velocities relative to their centers.

We're just referring to different coordinate systems whose origins coincide at a given point in spacetime, and describing how the light rays pass through that point in terms of the different coordinate systems at that instant. In this context the acceleration (or spinning) of the systems make no difference to the answer.

In other words, as long as our inertial coordinate system has the same velocity and orientation as the (ideal point-like) observer at the moment of the observation, it doesn't matter if the observer is in the process of changing his orientation or velocity. (This is a corollary of the "clock hypothesis" of special relativity, which asserts that a traveler's time dilation at a given instant depends only on his velocity and not his acceleration at that instant).

$$\sigma = \frac{(a+d)^2}{(ad-bc)}$$

In general we can classify Mobius transformations (and the corresponding Lorentz transformations) according to their "squared trace", i.e., the quantity

This is also the "conjugacy parameter", i.e., two linear fractional transformations are conjugate if and only if they have the same value of σ. The different kinds of transformations are listed below:

$0 \le \sigma < 4$	elliptic
$\sigma = 4$	parabolic
$\sigma > 4$	hyperbolic
$\sigma < 0$ or not real	loxodromic

For example, the class of pure rotations are a special case of elliptic transformations, having the form

$$w \rightarrow \frac{aw+b}{-bw+\bar{a}} \quad \text{with} \quad a\bar{a}+b\bar{b}=1$$

where an overbar denotes complex conjugation. Also, it's not hard to show that the compositions of an arbitrary linear fractional transformation f(z) are cyclical with a period m if and only if $\sigma = 4\cos(2k\pi/m)^2$.

We've seen that the general finite transformation of the incoming null rays can be expressed naturally in the form of a finite Mobius transformation of the complex plane (under sterographic projection). This is a very simple algebraic operation, given by the function

$$f(z) = \frac{az+b}{cz+d}$$

for complex constants a, b, c, d. This generates the discrete sequence $f_1(z) = f(z)$, $f_2(z) = f(f(z))$, $f_3(z) = f(f(f(z)))$, and so on for all $f_n(z)$ where n is a positive integer. It's also possible to parameterize a Mobius transformation to give the corresponding infinitesimal generator, which can be applied to give "fractional iterations" such as $f_{1/2}(z)$, or more generally the continuously parameterized transformation $f_p(z)$ for any real (or even complex) value of p.

To accomplish this we must (in general) first map the discrete generator f(z) to a domain in which it has some convenient exponential form, then apply the pth-order transformation, and then map back to the original domain. There are several cases to consider, depending on the character of the discrete generator.

In the degenerate case when ad = bc with $c \neq 0$, the pth iterate of f(z) is simply the constant $f_p(z) = a/c$. On the other hand, if c = 0 and $a = d \neq 0$, then $f_p(z) = z + (b/d)p$. The third case is with c = 0 and $a \neq d$. The pth iterate of f(z) in this case is

$$f_p(z) = \left(\frac{a}{d}\right)^p \left(z - \frac{b}{d-a}\right) + \left(\frac{b}{d-a}\right)$$

Notice that the second and third cases are really linear transformations, since c = 0. The fourth case is with $c \neq 0$ and $(a+d)^2/(ad-bc) = 4$, which leads to the following closed form expression for the pth iterate

$$f_p(z) = \frac{1}{\left(\frac{2c}{d-a+2cz}\right) + \left(\frac{2c}{a+d}\right)} + \left(\frac{a-d}{2c}\right)$$

This corresponds to the case when the two fixed points of the Mobius transformation are co-incident. In this "parabolic" case, if a+d = 0 then the Mobius transformation reduces to the first case with ad-bc = 0.

Finally, in the most general case we have $c \neq 0$ and $(a+d)^2 / (ad-bc) \neq 4$, and the pth iterate of f(z) is given by

$$f_p(z) = \frac{A}{1 - K^p\left(1 - \frac{A}{z-B}\right)} + B$$

where

$$A = \pm\sqrt{\left(\frac{a-d}{c}\right)^2 + \frac{4b}{c}} \qquad B = \left(\frac{a-d}{2c}\right)^2 \mp \sqrt{\left(\frac{a-d}{2c}\right)^2 + \frac{b}{c}}$$

This is the general case with two distinct fixed points. (If a+d = 0 then $\sigma = 0$ and K = -1).

The parameters A and B are the coefficients of the linear transformation that maps real line to the locus of points with real part equal to 1/2. Notice that the pth composition of f satisfies the relation

$$\left(1 - \frac{A}{f_p(z) - B}\right) = K^p\left(1 - \frac{A}{z - B}\right)$$

so we have

$$h\left(f_p(z)\right) = K^p h(z)$$

where

$$h(z) = \pm\frac{a-(A+B)}{z-B}$$

Thus h(f(z)) = K h(z), which shows that f(z) is conjugate to the simple function Kz. Since A+B is the complex conjugate of B, we see that h(z) can be expressed as

$$h(z) = \pm\frac{2cz - \rho}{2cz - \bar{\rho}}$$

where

$$\rho = (a-d) \pm \sqrt{(ad-bc)(\sigma - 4)} \quad \bar{\rho} = (a-d) \mp \sqrt{(a-d) + 4bc}$$

This enables us to express the pth composition of any linear fractional transformation with two fixed points, and therefore any corresponding Lorentz transformation, in the form

$$f_p(z) = h^{-1}(Kph(z))$$

This shows that there is a particular oriented frame of reference, represented by h(z), with respect to which the relation between the oriented frames z and f(z) is purely exponential. (We must refer to *oriented* frames rather than merely frames because the Mobius transformation represented the effects of general orientation as well as velocity boost). To show explicitly how the action of fp(z) on the complex plane varies with p, consider the relatively simple linear fractional transformation f(z) with fixed points at 0 and 1 on the real axis, which implies A = 1 and B = 0. In parameterized form the pth composition of this transformation is of the form

$$f_p(z) = \frac{z}{K^p + (1-K^p)z} = \frac{1}{\left(\frac{1-z}{z}\right)K^p + 1}$$

for some complex constant K, and the similarity parameter for this transformation is $\sigma = (1+K)^2/K$. For any given K and complex initial value z = x + iy, let

$$K = Re^{ie} \qquad u_r = \frac{x}{x^2+y^2} - 1 \qquad u_i = -\frac{y}{x^2+y^2}$$

Then the real and imaginary components of $f_p(z)$ are given by

$$f_p(z)_{real} = \frac{1 + R^2\left(u_r\cos(p\theta) - u_i\sin(p\theta)\right)}{1 + R^{2p}\left(u_r^2 + u_i^2\right) + 2R^p\left(u_r\cos(p\theta) - u_i\sin(p\theta)\right)}$$

$$f_p(z)_{imag} = -\frac{R^2\left(u_r\cos(p\theta) - u_i\sin(p\theta)\right)}{1 + R^{2p}\left(u_r^2 + u_i^2\right) + 2R^p\left(u_r\cos(p\theta) - u_i\sin(p\theta)\right)}$$

This makes explicit how the action of $f_p(z)$ on the complex plane is entirely determined by the magnitude and phase angle of the constant K, which, as we saw previously, is given by

$$K = \frac{1 \pm \sqrt{1-(4/\sigma)}}{1 \mp \sqrt{1-(4/\sigma)}}$$

If a, b, c, d are all real, then σ is real, in which case either K is real (σ>4 or σ<0) or K is complex (0<σ<4) with a magnitude (norm) of 1.

However, if a, b, c, d are allowed to be complex, then K can be complex with a magnitude other than 1. Of course, if K is real, we can set R = K and θ = 0, so pθ = 0 for all p, and the above equations reduce to

$$x_n = \frac{1 + K^n u_r}{1 + K^{2n}\left(u_r^2 + u_i^2\right) + K^n u_r}$$

$$y_n = -\frac{K^n u_r}{1 + K^{2n}\left(u_r^2 + u_i^2\right) + K^n u_r}$$

Clearly the computational complexity of the continuous parameterized transformation exceeds that of the discrete transformation. This raises an interesting question, at least from a neo-Platonic perspective.

Is it possible that nature prefers the simplicity of the discrete form over the continuous? In other words, are all physically realizable Lorentz transformations actually discrete? If so, what determines the "size" of the minimum transformation? What is the "size" of a Mobius transformation? We know that every Mobius is conjugate to a pure exponential, and the effect of which is a rotation and a re-scaling. In addition, the conjugation itself may impose some kind of *size*. It's interesting that the elements now are not frames but differences between frames, including rotations.

Thus, rather than the ontological objects of consideration being events of the form x, y, z, t, or even coordinate systems, the objects are the transformations with complex coefficients a, b, c, d. This again introduces the octonion space, though restricted by the fact that there are only three (complex) degrees of freedom.

Chapter 5

The Sagnac Effect

If two pulses of light are sent in opposite directions around a stationary circular loop of radius R, they will traveled the same inertial distance at the same speed, so they will arrive at the end point simultaneously. This is illustrated in the left-hand figure below.

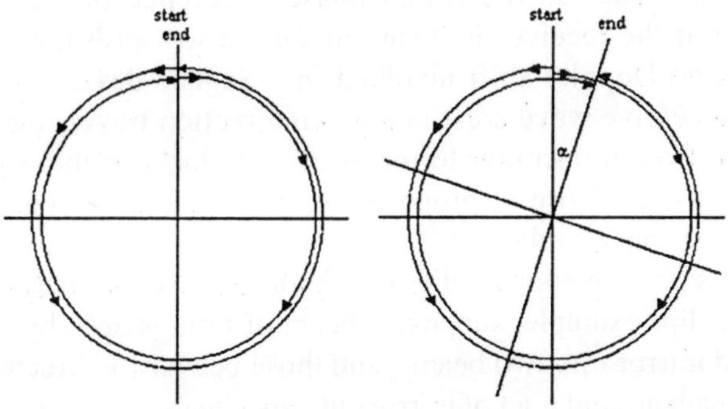

The figure on the right indicates what happens if the loop itself is rotating during this procedure. The symbol α denotes the angular displacement of the loop during the time required for the pulses to travel once around the loop. For any positive value of α, the pulse traveling in the same direction as the rotation of the loop must travel a slightly greater distance than the pulse traveling in the opposite direction. As a result, the counter-rotating pulse arrives at the "end" point slightly earlier than the co-rotating pulse.

Quantitatively, if we let ω denote the angular speed of the loop, then the circumferential tangent speed of the end point is $v = \omega R$, and the sum of the speeds of the wave front and the receiver at the "end" point is $c-v$ in the co-rotating direction and $c+v$ in the counter-rotating direction. Both pulses begin with an initial separation of $2\pi R$ from the end point, so the difference between the travel times is

$$\Delta t = 2\pi R\left(\frac{1}{c-v} - \frac{1}{c+v}\right) = \frac{4\pi R v}{c^2 - v^2} = \frac{4A\omega}{c^2 - v^2}$$

where $A = \pi R^2$ is the area enclosed by the loop. This analysis is perfectly valid in both the classical and the relativistic contexts. Of course, the result represents the time difference with respect to the axis-centered inertial frame. A clock attached to the perimeter of the ring would, according to special relativity, record a lesser time, by the factor $\gamma = (1-(v/c)^2)^{1/2}$, so the Sagnac delay with respect to such a clock would be $[4A\omega/c^2]/(1-(v/c)^2)^{1/2}$.

However, the characteristic frequency of a given light source co-moving with this clock would be greater, compared to its reduced value in terms of the axis-centered frame, by precisely the same factor, so the actual phase difference of the beams arriving at the receiver is invariant. (It's also worth noting that there is no Doppler shift involved in a Sagnac device, because each successive wave crest in a given direction travels the same distance from transmitter to receiver, and clocks at those points show the same lapse of proper time, both classically and in the context of special relativity).

This phenomenon applies to any closed loop, not necessarily circular. For example, suppose a beam of light is split by a half-silvered mirror into two beams, and those beams are directed in a square path around a set of mirrors in opposite directions as shown below.

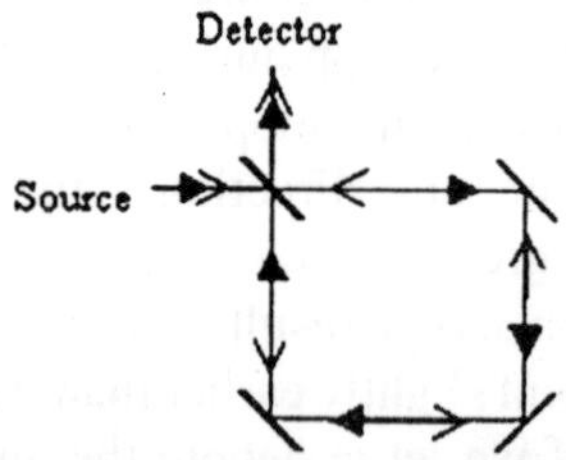

Just as in the case of the circular loop, if the apparatus is unaccelerated, the two beams will travel equal distances around the loop, and arrive at the detector simultaneously and in phase. However, if the entire device (including source and detector) is rotating, the beam traveling around the loop in the direction of rotation will have farther to go than the beam traveling counter to the direction of rotation, because during the period of travel the mirrors and detector will all move (slightly) toward the counter-rotating beam and away from the co-rotating beam. Consequently the beams will reach the detector at slightly different times, and slightly out of phase, producing optical interference "fringes" that can be observed and measured.

Michelson had proposed constructing such a device in 1904, but did not pursue it at the time, since he realized it would show only the absolute rotation of the device. The effect was first demonstrated in 1911 by Harress (unwittingly) and in 1913 by Georges Sagnac, who published two brief notes in the Comptes Rendus describing his apparatus and summarizing the results. He wrote

The result of measurements shows that, in ambient space, the light is propagated with a speed V_0, independent of the overall movement of the source of light O and optical system.

This rules out the ballistic theory of light propagation (as advocated by Ritz in 1909), according to which the speed of light is the vector sum of the velocity of the source plus a vector of magnitude c. Ironically, the original Michelson-Morley experiment was consistent with the ballistic theory, but inconsistent with the naove ether theory, whereas the Sagnac effect is consistent with the naove ether theory but inconsistent with the ballistic theory. Of course, both results are consistent with fully relativistic theories of Lorentz and Einstein, since according to both theories light is propagated at a speed independent of the state of motion of the source.

Because of the incredible precision of interferometric techniques, devices like this are capable of detecting and measuring extremely small amounts of absolute rotation. One of the first applications of this phenomenon was an experiment performed by Michelson and Gale in 1925 to measure the absolute rotation rate of the Earth by means of a rectangular optical loop 2/5 mile long and 1/5 mile wide. More recently, the invention of lasers around 1963 has led to practical small-scale devices for measuring rotation by exploiting the Sagnac effect.

There are two classes of such devices, namely, ring interometers and ring lasers. A *ring interferometer* typically consists of many windings of fibre optic lines, conducting light (of a fixed frequency) in opposite directions around a loop, and then recombining them to measure the phase difference, just as in the original Sagnac apparatus, but with greater efficiency and sensitivity.

A *ring laser*, on the other hand, consists of a laser cavity in the shape of a ring, which allows light to circulate in both directions, producing two standing waves with the same number of nodes in each direction. Since the optical path lengths in the two directions are different, the resonant frequencies of the two standing waves are also different. (In practice it is typically necessary to "dither" the ring to prevent phase locking of the two modes).

The "beat" between the two frequencies is measured, giving a result proportional to the rotation rate of the device. Incidentally, it isn't necessary for the actual laser cavity to circumscribe the entire loop; longitudinal pumping can be used, driven by feedback carried in opposite directions around the loop in ordinary optical fibers. (Needless to say, the difference in resonant frequency of the two stand waves in a ring laser due to the different optical path lengths is not to be confused with a Doppler shift). Today such devices are routinely used in guidance and navigation systems for commercial airliners, nautical ships, spacecraft, and in many other applications, and are capable of detecting rotation rates as slight as 0.00001 degree per hour.

We saw previously that the time delay (and therefore the difference in the optical path lengths) for a circular loop is proportional to the area enclosed by the loop. This interesting fact actually applies to arbitrary closed loops. To prove this, we will derive the difference in arrival times of the two pulses of light for an arbitrary polygonal loop inscribed in a circle. Let the (inertial) coordinates of two consecutive mirrors separated by a subtended angle θ be

$$x_1 = R\cos(\omega t) \quad y_1 = R\sin(\omega t)$$
$$x_2 = R\cos(\omega t + \theta) \quad y_2 = R\sin(\omega t + \theta)$$

where ω is the angular velocity of the device. Since light rays travel along null intervals, we have $c^2(dt)^2 = (dx)^2 + (dy)^2$, so the coordinate time T required for a light pulse to travel from one mirror to the next in the forward and reverse directions satisfies the equations

$$\begin{aligned} c^2T^2 &= 2R^2[1 - \cos(\omega T \pm \theta)] \\ &= 2R^2[1 - \cos(\omega T)\cos(\theta) \mp \sin(\omega T)\sin(\theta)] \end{aligned}$$

Typically ωT is extremely small, i.e., the polygon doesn't rotate through a very large angle in the time it takes light to go from one mirror to the next, so we can expand these equations in ωT (up to second order) and collect powers of T to give the quadratic

$$[c^2 - R^2\omega^2\cos(\theta)]T^2 \pm [2R^2\omega\sin(\theta)]T - 2R^2[1 - \cos(\theta)] = 0$$

The two roots of this polynomial are the values of T, one positive and one negative, for the co-rotating and counter-rotating solutions, so the difference in the absolute times is the sum of these roots. Hence we have

$$T_{co} - T_{counter} = \frac{2R^2\omega\sin(\theta)}{c^2 - R^2\omega^2\cos(\theta)}$$

This is the net contribution of this edge to the total time increment. Recalling that the area of a regular n-sided polygon of radius R is $nR^2\sin(2\pi/n)/2$, the area of the triangle formed by the hub and the two mirrors is $R^2\sin(\theta)/2$. It follows that each edge of an arbitrary polygonal loop inscribed in a circle contributes $4A_i\omega/(c^2 - v^2\cos(\theta))$ to the total time discrepancy, where A_i is the area of the ith triangular slice of the loop and v = Rω is the tangential speed of the mirrors. Therefore, the total discrepancy in travel times for the co-rotating and counter-rotating beams around the entire loop is simply

$$T_{co} - T_{counter} = \frac{4A\omega}{c^2 - R^2\cos(\theta)}$$

where A is the total area enclosed in the loop. This applies to polygons with any number of sides, including the limiting case of circular fibre-optic loops with virtually infinitely many edges (where the "mirrors" are simply the inner reflective lining of the fibre-optic cable), in which case θ goes to zero and the denominator of the phase difference is simply $c^2 - v^2$. For realistic values of v (i.e., very small compared with c), the phase difference reduces to the well-known result $4A\omega/c^2$. It's worth noting that nothing in this derivation is unique to special relativity, because the Sagnac effect is a purely "classical" effect. The apparatus is set up as a differential device, so the relativistic effects apply equally in both directions, and hence the higher-order corrections of special relativity cancel out of the phase difference.

Despite the ease and clarity with which special relativity accounts for the Sagnac effect, one occasionally sees claims that this effect entails a *conflict* with the principles of special relativity. The usual claim is that the Sagnac effect somehow falsifies the invariance of light speed with respect to all inertial coordinate systems.

Of course, it does no such thing, as is obvious from the fact that the simple description of an arbitrary Sagnac device given above is based on isotropic light speed with respect to one particular system of inertial coordinates, and all other inertial coordinate systems are related to this one by Lorentz transformations, which are defined as the transformations that preserve light speed. Hence no description of a Sagnac device in terms of any system of inertial coordinates can possibly entail non-isotropic light speed, nor can any such description yield physically observable results different from those derived above (which are known to agree with experiment).

Nevertheless, it remains a seminal tenet of anti-relativityism (for lack of a better term) that the trivial Sagnac effect somehow "disproves relativity". Those who espouse this view sometimes claim that the expressions "c+v" and "c–v" appearing in the derivation of the phase shift are *prima facie* proof that the speed of light is not c with respect to some inertial coordinate system.

When it is pointed out that those quantities do not refer to the speed of light, but rather to the sum and difference of the speed of light and the speed of some other object, both with respect to a single inertial coordinate system, which can be as great as 2c according to special relativity, the anti-relativityists are undaunted, and merely proceed to construct progressively more convoluted and specious "objections".

For example, they sometimes argue that each point on the perimeter of a rotating circular Sagnac device is always instantaneously at rest in *some* inertial coordinate system, and according to special relativity the speed of light is precisely c in all directions with respect to any inertial system of coordinates, so (they argue) the speed of light must be isotropic at every point around the entire circumference of the loop, and hence the light pulses must take an equal amount of time to traverse the loop in either direction.

Needless to say, this "reasoning" is invalid, because the pulses of light are never (let alone always) at the same point in the loop at the same time during their respective trips around the loop in opposite directions. At any given instant the point of the loop where one pulse is located is necessarily accelerating with respect to the instantaneous inertial rest frame of the point on the loop where the other pulse is located (and vice versa).

As noted above, it's self-evident that since the speed of light is isotropic with respect to at least one particular frame of reference, and since every other frame is related to that frame by a transformation that explicitly preserves light speed, no inconsistency with the invariance of the speed of light can arise.

Having accepted that the observable effects predicted by special relativity for a Sagnac device are correct and entail no logical inconsistency, the dedicated opponents of special relativity sometimes resort to claims that there is nevertheless an inconsistency in the relativistic *interpretation* of what's really happening locally around the device in certain extreme circumstances.

The fundamental fallacy underlying such claims is the idea that the beams of light are traveling the same, or at least congruent, inertial paths through space and time as they proceed from the source to the detector. If this were true, their inertial speeds would indeed need to differ in order for their arrival times at the detector to differ.

However, the two pulses do *not* traverse congruent paths from emission to detector (assuming the device is absolutely rotating).

The co-rotating beam is traveling slightly farther than the counter-rotating beam *in the inertial sense*, because the detector is moving away from the former and toward the latter while they are in transit. Naturally the ratio of optical path lengths is the same with respect to any fixed system of inertial coordinates.

It's also obvious that the absolute difference in optical path lengths cannot be "transformed away", e.g., by analyzing the process with respect to coordinates rigidly attached to and rotating along with the device. We can, of course, define a system of coordinates in terms of which the position of a point fixed on the disk is independent of the time coordinate, but such coordinates are necessarily rotating (accelerating), and special relativity does not entail invariant or isotropic light speed with respect to non-inertial coordinates. (In fact, one need only consider the distant stars circumnavigating the entire galaxy every 24 hours with respect to the Earth's rotating system of reference to realise that the limiting speed of travel is generally *not* invariant and isotropic in terms of accelerating coordinates).

A detailed analysis of a Sagnac device in terms of non-inertial (i.e., rotating) coordinates is presented in Section 4.8, and discussed from a different point of view in Section 5.1. For the present, let's confine our attention to inertial coordinates, and demonstrate how a Sagnac device is described in terms of instantaneously co-moving inertial frames of an arbitrary point on the perimeter.

Suppose we've sent a sequence of momentary pulses around the loop, at one-second intervals, in both directions, and we have photo-detectors on each mirror to detect when they are struck by a co-rotating or counter-rotating pulse. Clearly the pulses will strike each mirror at one-second intervals from both directions (though not necessarily synchronized) because if they were arriving more frequently from one direction than from the other, the secular lag between corresponding pulses would be constantly increasing, which we know is not the case.

So each mirror is receiving one pulse per second from both directions. Furthermore, a local measurement of light speed performed (over a sufficiently short period of time) by an observer riding along at a point on the perimeter will necessarily show the speed of light to be c in all direction with respect to his instantaneously co-moving inertial coordinates. However, this system of coordinates is co-moving with only one particular point on the rim. At other points on the rim these coordinates are not co-moving, and so the speed of light is not c at other points on the rim with respect to these coordinates.

To describe this in detail, let's first analyse the Sagnac device from the hub-centered inertial frame. Throughout this discussion we assume an n-sided polygonal loop where n is very large, so the segment between any two adjacent mirrors subtends only a very small angle. With respect to the hub-centered frame each segment is moving with a velocity v parallel to the direction of travel of the light beams, so the situation on each segment is as plotted below in terms of hub-frame coordinates:

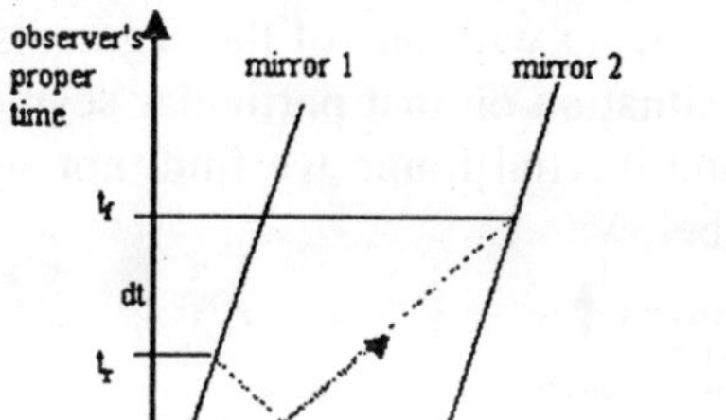

In this drawing, t_f is the time required for light to cross this segment in the co-rotating direction, and t_r is the time required for light to cross in the counter-rotating direction. The difference between these two times, denoted by dt, is the incremental Sagnac effect for a segment of length dp on the perimeter.

Now, the ratio of dt/dp as a function of the rim velocity v can easily be read off this diagram, and we find that

$$\frac{dt}{dp} = \frac{2v}{c^2 - v^2}$$

This can be taken as a measure of the anisotropy over an incremental segment with respect to the hub frame. (Notice that this anisotropy with respect to the conventional relativistic spacetime decomposition for any inertial frame is actually in the distance traveled, not the speed of travel). All the segments are symmetrical in this frame, so they all have this same anisotropy. Therefore, we can determine the total difference in travel times for co-rotating and counter-rotating beams of light making a complete trip around the loop by integrating dt around the perimeter. Thus we have

$$T = \int_{p=0}^{2\pi r} \frac{2v}{c^2 - v^2} dp = \frac{4\pi r v}{c^2 - v^2}$$

Substituting ωr in place of v in the numerator, and noting that the enclosed area is $A = \pi r^2$, we again arrive at the result $T = 4A\omega/(c^2 - v^2)$.

Now let's analyse the loop with respect to one of our tangential frames of reference, i.e., an inertial frame that is momentarily co-moving with one of the segments on the rim. If we examine the situation on that particular segment in terms of its own co-moving inertial frame we find, not surprisingly, the situation shown below:

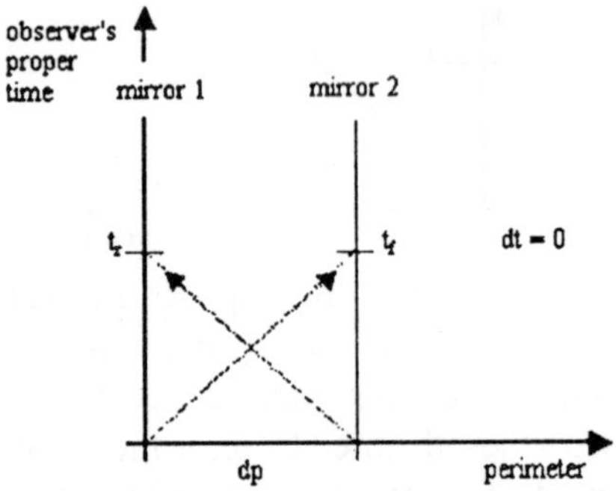

This shows that dt/dp = 0, meaning no anisotropy at all. Nevertheless, if the light beams are allowed to go all the way around the loop, their total travel times will differ by T as computed above, so how does that difference arise with respect to this tangential frame?

Notice that although dt/dp equals zero at this tangent point with respect to the tangent frame, segments 90 degrees away from this point have the same anisotropy as we found for all the segments relative to the hub frame, namely, $dt/dp = 2v/(c^2 - v^2)$, because the velocity of those two segments relative to our tangential frame is exactly v along the direction of the light rays, just as it was with respect to the hub frame. Furthermore, the segment 180 degrees away from our tangent segment has *twice* the anisotropy as it has with respect to the original hub-frame inertial coordinates, because that segment has a velocity of 2v with respect to our tangential frame.

In general, the anisotropy dt/dp can be computed for any segment on the loop simply by determining the projection of that segment's velocity (with respect our tangential frame) onto the axis of the light rays. This gives the results illustrated below, showing the ratio of the tangential frame anisotropy to the hub frame anisotropy:

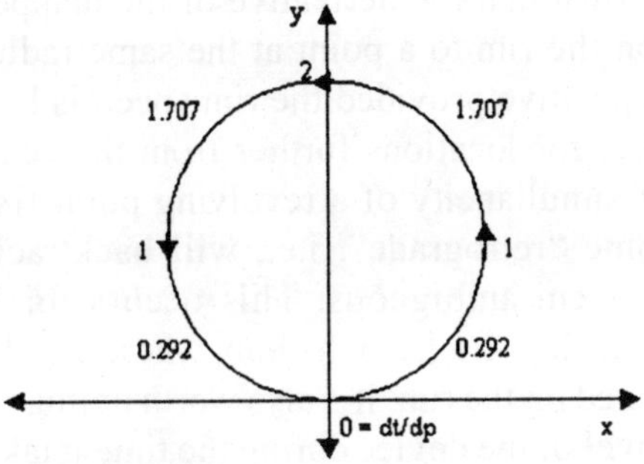

It's easy to show that

$$\frac{dt}{dp} = \frac{2v}{c^2 - v^2}[1 - \cos(\theta)]$$

where θ is the angle relative to the tangent point. To assess the total difference in arrival times for light rays going around the loop in opposite directions, we need to integrate dt by dp around the perimeter. Noting that θ equals p/r, we have

$$T = \int_{p=0}^{2\pi r} \frac{2v[1 - \cos(p/r)]}{c^2 - v^2}\, dp = \frac{4\pi rv}{c^2 - v^2}$$

which again equals $4A\omega/(c^2 - v^2)$, in agreement with the hub frame analysis. Thus, although the anisotropy is zero at each point on the rim's surface when evaluated with respect to that point's co-moving inertial frame, we always arrive at the same overall non-zero anisotropy for the entire loop. This was to be expected, because the absolute physical situation and intervals are the same for all inertial frames. We're simply decomposing those absolute intervals into space and time components in different ways.

The union of all the "present" time slices of the sequence of instantaneous co-moving inertial coordinate systems for a point fixed on the rim of a rotating disk, with each time slice assigned a time coordinate equal to the proper time of the fixed point, constitutes a coherent and unambiguous coordinate system over a region of spacetime that includes the entire perimeter of the disk. The general relation for mapping the proper time of one worldline into another by means of the co-moving planes of simultaneity of the former is derived at the end of Section 2.9, where it is shown that the derivative of the mapped time from a point fixed on the rim to a point at the same radius fixed in the hub frame is positive provided the rim speed is less than c.

Of course, for locations further from the centre of rotation the planes of simultaneity of a revolving point fixed on the rim will be become "retrograde", i.e., will backtrack, making the coordinate system ambiguous. This occurs for locations at a distance greater than 1/a from the hub, where a is the acceleration of the point fixed on the rim. It's also worth noting that the amount of angular travel of the device during the time it takes for one pair of light pulses to circumnavigate a circular loop is directly proportional to the net "anisotropy" in the travel times.

To prove this, note that in a circular Sagnac device of radius R the beam of light in the direction of rotation travels a distance of $(2\pi - \omega t_1)R$ and the other beam goes a distance of $(2\pi + \omega t_2)R$ where t_1 and t_2 are the travel times of the two beams, and ω is the angular velocity of the loop.

The travel times of the beams are just these distances divided by c, so we have

$$ct_1 = (2\pi - \omega t_1) \qquad ct_2 = (2\pi + \omega t_2)$$

Solving for the times gives

$$t_1 = \frac{2\pi R}{c + \omega R} \qquad t_2 = \frac{2\pi R}{c + \omega R}$$

so the difference in times is

$$t_2 - t_1 = \frac{4A\omega}{c^2 - v^2}$$

where $A = 2\pi R^2$ and $v = \omega R$. The "anisotropic ratio" is the ratio of the travel times, which is

$$\frac{t_2}{t_1} = \frac{c + \omega R}{c - \omega R}$$

Solving this for ωR gives

$$\omega R = c\left(\frac{t_2 / t_1 - 1}{t_2 / t_1 + 1}\right)$$

Letting θ denote the angular travel of the loop during the travel of the two light beams, we have

$$\theta = \omega t_2 = \frac{2\pi\omega R}{c - \omega R}$$

Substituting for ωR this reduces to

$$\theta = \pi\left(\frac{t_2}{t_1} - 1\right).$$

Therefore, the amount by which the ratio of travel times differs from 1 is exactly proportional to the angle through which the loop rotates during the transit of light, and this is true independent of R. (Of course, increasing the radius has the effect of increasing the *difference* between the travel times, but it doesn't alter the ratio).

It's worth emphasizing that the Sagnac effect is purely a classical, not a relativistic phenomenon, because it's a "differential device", i.e., by running the light rays around the loop in opposite directions and measuring the time difference, it effectively cancels out the "transverse" effects characteristic of truly relativistic phenomenon.

For example, the length of each incremental segment around the perimeter is shorter by a factor of $[1-(v/c)^2]^{1/2}$ in the hub based frame than in it's co-moving tangential frame, but this factor applies in both directions around the loop, so it doesn't affect the differential time. Likewise a clock on the perimeter moving at the speed v runs slow, in accord with special relativity, but the frequency of the light source is correspondingly slow, and this applies equally in both directions, so this does not affect the phase difference at the receiver.

Thus, a pure Sagnac apparatus does not discriminate between relativistic and pre-relativistic theories (although it does rule out ballistic theories, ala Ritz). Ironically, this is the main reason it comes up so often in discussions of relativity, because the effect can easily be computed on a non-relativistic basis (as we did above for a circular loop, taking the sums c+v and c–v to determine the transit times in the two directions).

Of course, if the light traveling around the loop passes through moving media with indices of refraction differing significantly from unity, then the Fizeau effect must also be taken into account, and in this case the results, while again perfectly consistent with special relativity, are quite problematic for any non-relativistic ether-based interpretation.

As mentioned above, as early as 1904 Michelson had proposed using such a device to measure the rotation of the earth, but he hadn't pursued the idea, since measurements of absolute rotation are fairly commonplace (e.g. Focault's pendulum). Nevertheless, he (along with Gale) agreed to perform the experiment in 1925 (at considerable cost) at the urging of "relativists", who wished him to verify the shift of 236/1000 of a fringe predicted by special relativity. This was intended mainly to refute the ballistic theory of light propagation, which predicts zero phase shift (for a circular device). Michelson was not enthusiastic, since classical optics on the assumption of a stationary ether predicted exactly the same shift does special relativity (as explained above). He said

We will undertake this, although my conviction is strong that we shall prove only that the earth rotates on its axis, a conclusion which I think we may be said to be sure of already.

As Harvey Lemon wrote in his biographical sketch of Michelson, "The experiment, performed on the prairies west of Chicago, showed a displacement of 230/1000, in very close agreement with the prediction. The rotation of the Earth received another independent proof, the theory of relativity another verification. But neither fact had much significance." Michelson himself wrote that "this result may be considered as an additional evidence in favour of relativity - or equally as evidence of a stationary ether".

The only significance of the Sagnac effect for special relativity (aside from providing another refutation of ballistic theories) is that although the effect itself is of the first order in v/c, the qualitative description of the local conditions on the disk in terms of inertial coordinates depends on second-order effects.

These effects have been confirmed empirically by, for example, the Michelson-Morley experiment. Considering the Earth as a particle on a large Sagnac device as it orbits around the Sun, the ether drift experiments demonstrate these second-order effects, confirming that the speed of light is indeed invariant in terms of relatively moving systems of inertial coordinates.

REFRACTION AT A PLANE BOUNDARY BETWEEN MOVING MEDIA

The ray angles θ_1 and θ_2 for incident and refracted optical rays at a plane boundary between regions of constant indices of refraction n_1 and n_2 are related according to Snell's law

$$n_1 \sin(\theta_1) = n_2 \sin(\theta_2)$$

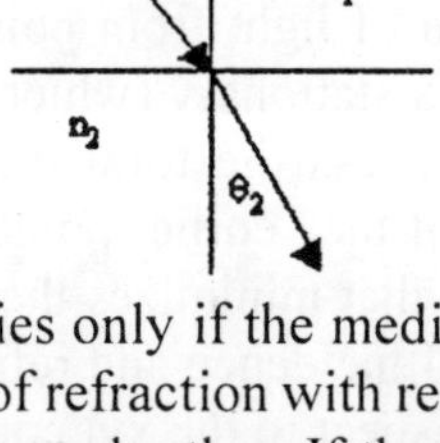

However, this formula applies only if the media (which are assumed to have isotropic index of refraction with respect to their rest frames) are at rest relative to each other. If the media are in relative transverse motion, it is necessary to account for the effect of aberration on the ray angles relative to the rest frames of the respective media.

The result is that the effective refraction is a function of the relative transverse velocity of the media. Thus, measurements of the optical refraction could (in principle) be used to determine the velocity of a moving volume of fluid. Unlike Doppler shift measurement techniques, this approach does not rely on the presence of discrete particles in the fluid, and involves only measurements of direct, rather than reflected, light signals.

Since the amount of refraction at a boundary depends on the angle of incidence with respect to the rest frames of the media, it follows that if the media have different rest frames the simple form of Snell's law does not apply directly, because it will be necessary to account for aberration. To derive the law of refraction for transversely moving media, consider the arrangement shown in Figure 1, drawn with respect to a system of coordinates (x, y, t) relative to which the medium with refractive index n_1 is at rest.

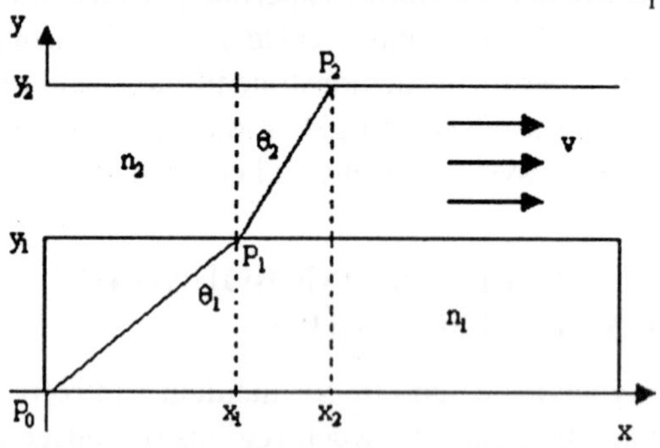

In these coordinates the medium with index n_2 is moving transversely with a speed v. By both Fermat's principle of "least time" and the principles of quantum electrodynamics, we know that the path of light from point P_0 to point P_2 is such that the travel time is stationary (which, in this case, means minimized), so if we express the total travel time as a function of the x coordinate of the "corner point" P_1, we can differentiate to find the position that minimizes the time, and from this we can infer the angles of incidence and refraction.

With respect to the xyt coordinates in which the n_1 medium is at rest, the squared spatial distance from P_0 to P_1 is $x_1^2 + y_1^2$, so the time required for light to traverse that distance is

$$t_1 - t_0 = n_1\sqrt{x_1^2 + y_1^2}$$

On the other hand, for the trip from point P_1 to point P_2 we need to know the distance traveled with respect to the coordinates x'y't' in which the n_2 medium is at rest. If we define

$$\Delta x = x_2 - x_1 \quad \Delta y = y_2 - y_1 \quad \Delta t = t_2 - t_1$$

then the Lorentz transformation gives us the corresponding increments in the primed coordinates

$$\Delta x' = \frac{\Delta x - v\Delta t}{\sqrt{1-v^2}} \quad \Delta y' = \Delta y \quad \Delta t' = \frac{\Delta t - v\,\Delta x}{\sqrt{1-v^2}}$$

Therefore, the squared spatial and temporal distances from P_1 to P_2 in the n_2 rest coordinates are given by

$$(\Delta x')^2 + (\Delta y')^2 = \frac{\Delta x^2 - 2v\,\Delta x\,\Delta t + v^2\Delta t^2}{1-v^2} + \Delta y^2$$

$$(\Delta t')^2 = \frac{\Delta t^2 - 2v\,\Delta x\,\Delta t + v^2\Delta x^2}{1-v^2}$$

Since the ratio of these increments equals the square of the speed of light in the n_2 medium, which is $1/n_2^2$, we have

$$\Delta t^2 - 2v\,\Delta x\,\Delta t + v^2\Delta x^2 = n_2^2$$

$$\left[\Delta x^2 + \Delta y^2 - 2v\,\Delta x\,\Delta t + v^2\left(\Delta t^2 + \Delta y^2\right)\right]$$

Solving this quadratic for Δt, which equals $t_C - t_B$, gives

$$\Delta t = \frac{2v\Delta x\left(1-n^2\right) - n\sqrt{1-v^2}\sqrt{\left(1-v^2\right)\Delta x^2 + \left(1-n^2v^2\right)\Delta y^2}}{1-n^2v^2}$$

Differentiating with respect to Δx, and noting that $d(\Delta x)/dx_1 = -1$, we can minimize the total travel time $t_2 - t_0$ by adding the derivatives of Δt and $t_1 - t_0$ with respect to x_1, and setting the result to zero. This leads to the condition

$$\left(1-n_2^2v^2\right)n_1\frac{x_1}{\sqrt{x_1^2+y_1^2}}$$

$$-\frac{\left(1-v^2\right)^{3/2} n_2\Delta x}{\sqrt{\Delta x^2 + \Delta y^2 - v^2\left(\Delta x^2 + n_2^2\Delta y^2\right)}} = v\left(1-n_2^2\right)$$

Making the substitutions

$$\sin(\theta_1) = \frac{x_1}{\sqrt{x_1 + y_1}}$$

$$\sin(\theta_2) = \frac{\Delta x}{\sqrt{\Delta x^2 + \Delta y^2}} \quad \cos(\theta_2) = \frac{\Delta y}{\sqrt{\Delta x^2 + \Delta y^2}}$$

we arrive at the equation for refraction at the plane boundary between transversely moving media

$$\left[1 + v^2\left(\frac{1 - n_2^{\,2}}{1 - v^2}\right)\right] n_1 \sin(\theta_1)$$

$$= \frac{n_2 \sin(\theta_2)}{\sqrt{1 + v^2 \cos^2(\theta_2)\left(\frac{1 - n_2^{\,2}}{1 - v^2}\right)}} + v\left(\frac{1 - n_2^{\,2}}{1 - v^2}\right)$$

As expected, this reduces to Snell's law for stationary media if we set $v = 0$. Also, if the moving medium has a refractive index of $n_2 = 1$, this equation again reduces to Snell's law, regardless of the velocity, because the concept of speed doesn't apply to the vacuum.

If we define the parameter

$$\Psi_i = \frac{1 - n_i^{\,2}}{1 - v^2}$$

then the refraction equation can be written more compactly as

$$\left[1 + v^2\Psi_2\right] n_1 \sin(\theta_1) = \frac{n_2 \sin(\theta_2)}{\sqrt{1 + v^2 \cos^2(\theta_2)\Psi_2}} + v\Psi_2$$

This can be solved explicitly for $\sin(\theta_2)$ to give the result

$$\sin(\theta_2) = \sqrt{\frac{1 + \Psi_2 v^2}{\Psi_2 v^2 + \left(\frac{n_2}{\left(1 + \Psi_2 v^2\right) n_1 \sin(\theta_1) - \Psi_2 v}\right)^2}}$$

with the appropriate sign for the square root. Taking $n_1 = 1.2$ and

$n_2 = 1.5$, the figure below shows the angle of refraction θ_2 as a function of the transverse speed v of the medium with various angles of incidence θ_1 ranging from $-3\pi/8$ to $+3\pi/8$ radians.

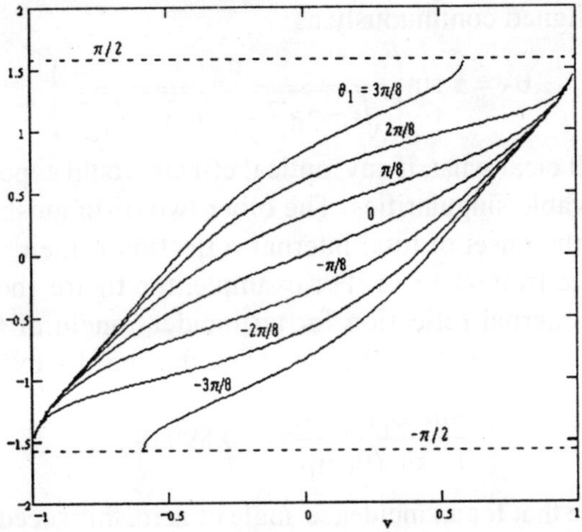

Incidentally, when plotting these lines it is necessary to take the positive root when v is above the zero-crossing speed, and the negative root when v is below. The zero-crossing speed (i.e., the speed v when the refracted angle is zero) is

$$v_0 = \frac{\left(n_2^{\ 2}-1\right)-\sqrt{\left(n_2^{\ 2}-1\right)^2+\left[2n_1n_2\sin(\theta)\right]^2}}{\left[2n_1n_2\sin(\theta)\right]n_2}$$

The figure shows that at high relative speeds and high angle of incidence we can achieve total internal reflection, even though the downstream medium is more dense than the upstream medium. The critical conditions occur when the squared quantity in parentheses in the preceding equation reaches ± 1, which implies

$$\left(1+\Psi_2 v^2\right)n_1\sin(\theta_1)-\Psi_2 v = \pm\, n_2$$

Solving these two quadratics for v (remembering that ψ_2 is a function of v), we have the four distinguished speeds

$$\frac{1}{n_2} \quad -\frac{1}{n_2} \quad \frac{\sin(\theta_1)n_1 - n_2}{1-\sin(\theta_1)n_1n_2} \quad \frac{\sin(\theta_1)n_1 + n_2}{1-\sin(\theta_1)n_1n_2}$$

The two speeds given by $\pm 1/n_2$ (which are just the speeds of light in the moving medium) generally correspond to removable singularities, because both the numerator and denominator of the expression for $\sin(\theta_2)$ vanish. At these speeds the values of θ_2 can be assigned continuously as

$$\theta_2 = a\ \sin\left(\frac{n_2}{\sqrt{1 \mp 2n_1{}^2 \sin(\theta_1) + 2n_2{}^2}}\right)$$

It isn't clear what, if any, optical effects would appear at these two removable singularities. The other two distinguished speeds represent the onset of total internal reflection if their values fall in the range from -1 to +1. For example, the figure above shows that total internal reflection for an incident angle of $\theta_1 = 3\pi/8$ with $n_1 = 1.2$ and $n_2 = 1.5$ begins when the speed v exceeds

$$\frac{\sin(\theta_1)n_1 - n_2}{1 - \sin(\theta_1)n_1 n_2} = 0.59028...$$

Notice that for an incidence angle of zero, this speed is simply n_2, which is ordinarily greater than 1, and thus outside the range of achievable speeds (since we assume the medium itself is moving through a vacuum).

However, for non-zero angles of incidence it is possible for one of these two critical speeds to lie in the achievable range. In fact, for certain values of n_1, n_2, and θ_1, it is possible for all four of the critical speeds to lie within the achievable range, leading to some interesting phenomena. For example, with $n_1 = n_2 = 2.5$ and with $\theta_1 = 45$ degrees, the refracted angle as a function of medium speed is as shown below.

In this case the distinguished speeds are -0.4, +0.203, +0.4, and +0.783. This suggests that as the transverse speed of the medium increases from 0, the refracted ray becomes steeper until reaching 90 degrees at v = +0.203, at which point there is total internal reflection. This remains the case until achieving a speed of +0.783, at which point some refraction is re-introduced, and the refracted angle sweeps back from +90 to about +80 degrees (relative to the stationary frame), and then back to +90 degrees as speed continues to increase to 1. This can be explained in terms of the variations in the effective critical angle and the aberration angle. As speed increases, the effective critical angle for total internal reflection initially increases faster than the aberration angle, pushing the ray into total internal reflection.

However, eventually (at close to the speed of light) the aberration effect brings the incident ray back into the refractive range.

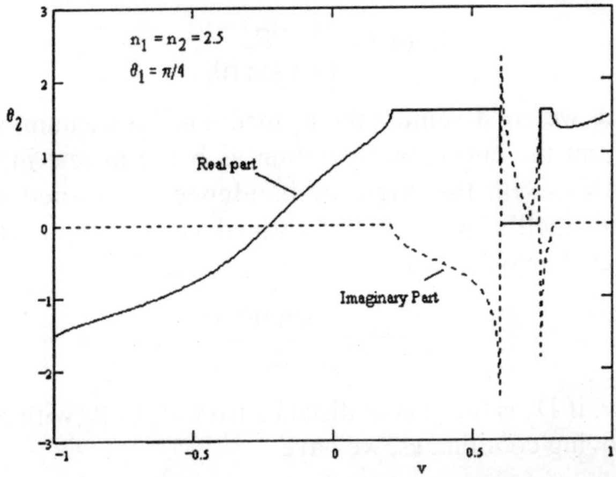

For an alternative derivation that leads to a different, but equivalent, relation, suppose the index of refraction of the stationary region is $n_1 = 1$, which implies this region is a vacuum. If we let d_1 denote the spatial distance from P_0 to P_1 with respect to the rest frame, then we have

$$x_1 = d_1 \sin(\theta_1) \quad y_1 = d_1 \cos(\theta_1) \quad t_1 = d_1$$

These are the components of the interval P_0 to P_1 with respect to the rest frame of n_1, and they can be converted to the frame of n_2 (denoted by upper case letters) using the Lorentz transformation

$$X_1 = \frac{x_1 - vt_1}{\sqrt{1-v^2}} = \frac{d_1 \sin(\theta_1) + vd_1}{\sqrt{1-v^2}}$$

$$X_1 = y_1 = d_1 \cos(\theta_1)$$

$$T_1 = \frac{t_1 - vx_1}{\sqrt{1-v^2}}$$

Letting Θ_1 denote the angle θ_1 with respect to the moving n_2 coordinate system, we can express the tangent of this angle as

$$\tan(\Theta_1) = \frac{X_1}{Y_1} = \frac{\sin(\theta_1) + v}{\cos(\theta_1)\sqrt{1-v^2}}$$

Taking the sine of the inverse tangent of both sides gives the familiar aberration formula

$$\sin(\Theta_1) = \frac{\sin(\theta_1) + v}{1 + v\sin(\theta_1)}$$

Since we are assuming the n_1 medium is a vacuum, we are free to treat the entire configuration as being at rest in the n_2 coordinates, with the angle of incidence as defined above. Therefore, Snell's law for stationary media can be applied to give the refracted angle relative to these coordinates

$$\sin(\Theta_2) = \frac{\sin(\Theta_1)}{n_2}$$

Now, if D_2 is the spatial distance from P_1 to P_2 with respect to the moving coordinates, we have

$$X_2 - X_1 = D_2\sin(\Theta_2) \quad Y_2 - Y_1 = D_2\cos(\Theta_2) \quad T_2T_1 = n_2D_2$$

Also, the Lorentz transformation gives the coordinates of points P_1 and P_2 in the rest frame in terms of the coordinates in the moving frame as follows:

$$x_1 = \frac{X_1 - vT_1}{\sqrt{1-v^2}} \quad x_2 = \frac{X_2 - vT_2}{\sqrt{1-v^2}} \quad y_2 = Y_2$$

From these we can construct the tangent of θ_2 with respect to the rest coordinates

$$\tan(\theta_2) = \frac{x_2 - x_1}{y_2 - y_1} = \frac{(X_1 - vT_2) - (X_1 - vT_1)}{\sqrt{1-v^2}(Y_2 - Y_1)}$$

$$= \frac{(X_2 - X_2)\text{-}v(T_2 - T_1)}{\sqrt{1-v^2}(Y_2 - Y_1)}$$

Substituting for the coordinate differences gives

$$\tan(\theta_2) = \frac{\sin(\Theta_2) - n_2v}{\cos(\Theta_2)\sqrt{1-v^2}}$$

We saw previously that

$$\sin(\Theta_2) = \frac{1}{n_2}\left(\frac{\sin(\theta_1) + v}{1 + v\sin(\theta_1)}\right)$$

so we can explicitly compute θ_2 from θ_1. It can be shown that this solution is identical to the solution (with $n_1 = 1$) derived previously on the basis of Fermat's principle. Furthermore, we can solve these equations for $\sin(\theta_1)$ as a function of θ_2 and then by equating this $\sin(\theta_1)$ with $n_3 \sin(\theta_3)$ for a stationary medium neighboring the vacuum region, we again have the general solution for two refractive media in relative transverse motion. A plot of θ_2 from θ_1 for various values of v is shown below:

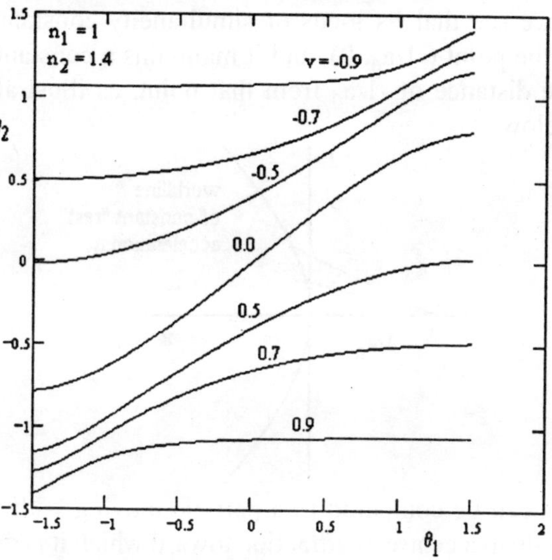

ACCELERATED TRAVELS

Suppose a particle accelerates in such a way that it is subjected to a constant *proper* acceleration a_0 for some period of time. The proper acceleration of a particle is defined as the acceleration with respect to the particle's momentarily co-moving inertial coordinates at any given instant.

The particle's velocity is v = 0 at the time t = 0, when it is located at x = 0, and at some infinitesimal time Δt later its velocity is $\Delta t\ a_0$ and its location is $(1/2)\ a_0\ \Delta t^2$. The slope of its line of simultaneity is the inverse of the slope 1/v of its worldline, so its locus of simultaneity at t = Δt is the line given by

$$(t - \Delta t) = \Delta t\ a_0 \left(x - \frac{1}{2} a_0 \Delta t^2 \right)$$

This line intersects the particle's original locus of simultaneity at the point (x, 0) where

$$x = -\frac{1}{a_0} + \frac{1}{2} a_0 \Delta t^2$$

At each instant the particle is accelerating relative to its current instantaneous frame of reference, so in the limit as Δt goes to zero we see that its locus of simultaneity constantly passes through the point $(-1/a_0, 0)$, and it maintains a constant absolute spacelike distance of $-1/a_0$ from that point, as illustrated in the figure below.

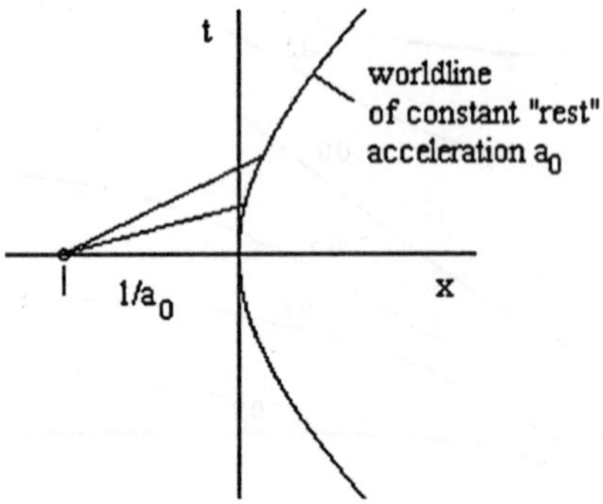

This can be compared to a particle moving with a speed v tangentially to a centre of attraction toward which it is drawn with a constant acceleration a_0. The path of such a particle is a circle in space of radius v^2/a_0. Likewise in spacetime a particle moving with a speed c tangentially to a centre of "repulsion" with a constant acceleration a_0 traces out a hyperbola with a "radius" of c^2/a_0. (In this discussion we are using units with c=1, so the "radius" shown in the above figure is written as $1/a_0$).

Since the worldline of a particle with constant proper acceleration is a branch of a hyperbola with "radius" $1/a_0$, we can shift the x axis by $1/a_0$ to place the origin at the centre of the hyperbola, and then write the equation of the worldline as

$$x^2 - t^2 = \frac{1}{a_0{}^2}$$

Differentiating both sides with respect to t gives

$$2x\frac{dx}{dt} - 2t = 0$$

which shows that the velocity of the worldline at any point (x, t) is given by v = t/x. Consequently the line from the origin through any point on the hyperbolic path represents the space axis for the co-moving inertial coordinates of the accelerating worldline at that point.

The same applies to any other hyperbolic path asymptotic to the same lightlines, so a line from the origin intersects any two such hyperbolas at points that are mutually simultaneous and separated by a constant proper distance (since they are both a fixed proper distance from the origin along their mutual space axis). It follows that in order for a slender "rigid" rod accelerating along its axis to maintain a constant proper length (with respect to its co-moving inertial frames), the parts of the rod must accelerate along a family of hyperbolas asymptotic to the same lightlines, as illustrated below.

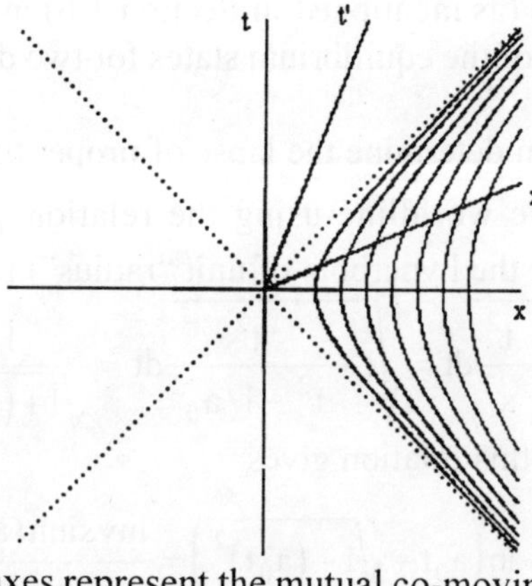

The x', t' axes represent the mutual co-moving inertial frame of the hyperbolic worldlines where they intersect with the x' axis. All the worldlines have constant proper distances from each other along this axis, and all have the same speed. The latter implies that they have each been accelerated by the same total amount at any instant of their mutual co-moving inertial frame, but the accelerations have been distributed differently.

The "inner-most" worldline (i.e., the trailing end of the rod) has been subjected to a higher level of instantaneous acceleration but for a shorter time, whereas the "outer-most" worldline (i.e., the leading end of the rod) has been accelerated more mildly, but for a longer time.

It's worth noting that this form of "coherent" acceleration would not occur if the rod were accelerated simply by pushing on one end. It would require the precisely coordinated application of distinct force profiles to each individual particle of the rod. Any deviation from these profiles would result in internal stresses of one part of the rod on another, and hence the rest length would not remain fixed.

Furthermore, even if the coherent acceleration profiles are perfectly applied, there is still a sense in which the rod has not remained in complete physical equilibrium, because the elapsed proper times along the different hyperbolic worldlines as the rod is accelerated from a rest state in x, t to a rest state in some x', t' differ, and hence the quantum phases of the two ends of the rod are shifted with respect to each other. Thus we must assume memorylessness (as mentioned in Section 1.6) in order to assert the equivalence of the equilibrium states for two different frames of reference.

We can then determine the lapse of proper time τ along any given hyperbolic worldline using the relation $d\tau = dt\sqrt{1-v^2}$, which leads (for the hyperbola of unit "radius") to

$$d\tau = \sqrt{1-\frac{t^2}{x^2}}dt = \sqrt{1-\frac{t^2}{t^2+1/{a_0}^2}}dt = \frac{1}{\sqrt{1+(a_0 t)}}dt$$

Integrating this relation gives

$$\tau = \frac{1}{a_0}\ln\left(a_0 t + \sqrt{1+(a_0 t)^2}\right) = \frac{\operatorname{invsinh}(a_0 t)}{a_0}$$

Solving this for t and substituting into the equation of the hyperbola to give x, we have the parametric equation of the hyperbola as a function of the proper time along the worldline. If we subtract $1/a_0$ from x to return to our original x coordinate (such that x = 0 at t = 0) these equations are

$$x = \frac{\cosh(a_0\tau)}{a_0} \qquad t = \frac{\sinh(a_0\tau)}{a_0}$$

Differentiating the above expressions gives

$$\frac{dx}{d\tau} = \sinh(a_0\tau) \qquad \frac{dt}{d\tau} = \cosh(a_0\tau)$$

so the particle's velocity relative to the original inertial coordinates is

$$\frac{dx}{dt} = \left(\frac{d\tau}{dt}\right) = \tanh(a_0\tau)$$

We're using "time units" throughout this section, which means that all times and distances are expressed in units of time. For example, if the proper acceleration of the particle is 1g (the acceleration of gravity at the Earth's surface), then

$$g = (3.27)10^{-8}\ \text{sec}^{-1} = 1.031\ \text{years}^{-1}$$

and all distances are in units of light-seconds.

To show the implications of these formulas, suppose a space traveller moves away from the Earth with a constant proper acceleration of 1g for a period of T years as measured on Earth. He then reverses his acceleration, coming to rest after another T years has passed on Earth, and then continues his constant Earthward acceleration for another T Earth-years, at which point he reverses his acceleration again and comes to rest back at the Earth in another T Earth-years. The total journey is completed in 4T Earth-years, and it consists of 4 similar hyperbolic segments as illustrated below.

There are several questions we might ask about this journey. First, how far away from Earth does the traveller reach at his furthest point?

This occurs at point C, which is at 2T according to Earth time, when the traveler's acceleration brings him momentarily to rest with respect to the Earth. To answer this question, recall that τ can be expressed as a function of t by

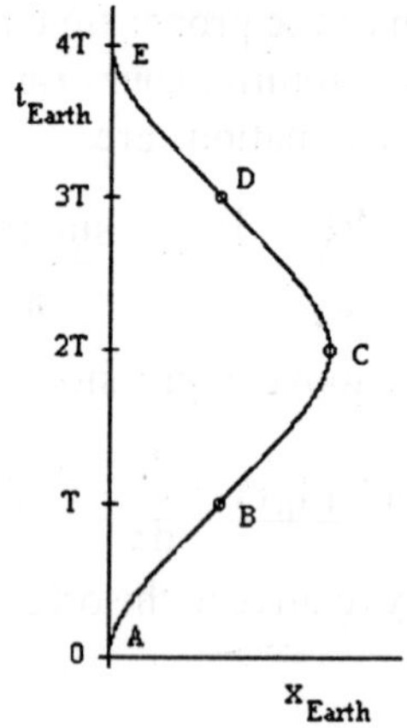

$$\tau = \frac{\text{inv}\sinh(gt)}{g}$$

Now, the maximum distance from Earth is twice the distance at point B, when t = T, so we have

$$x_c = 2\frac{\cosh((\text{inv}\sinh(dT)) - 1}{g} = 2\left(\sqrt{(1/g)^2 + T^2} - (1/g)\right)$$

The maximum speed of the traveller in terms of the Earth's inertial coordinates occurs at point B, where t = T (and again at point D, where t = 3T), and so is given by

$$v_B = \tanh(\text{inv}\sinh(gT))$$

The total elapsed proper time for the traveller during the entire journey out and back, which takes 4T years according to Earth time, is 4 times the lapse of proper time to point B at t = T, so it is given by

$$\tau = 4\frac{(\text{inv}\sinh(gT)}{g}$$

So far we have focused mainly on a description of events in terms of the Earth's inertial coordinates x and t, but we can also describe the same events in terms of coordinate systems associated with the accelerating traveller.

At any given instant the traveller is momentarily at rest with respect to a system of inertial coordinates, so we can define "proper" time and space measurements in terms of these coordinates. However, when we differentiate these time and space intervals as the traveller progresses along his worldline, we will find that new effects appear, due to the fact that the coordinate system itself is changing.

As the traveller accelerates he continuously progresses from one system of momentarily co-moving inertial coordinates to another, and the effect of this change in the coordinates will show up in any derivatives that we take with respect to the time and space components.

For example, suppose we ask how fast the Earth is moving relative to the traveller. This question can be interpreted in different ways. With respect to the traveler's momentarily co-moving inertial coordinates, the Earth's velocity is equal and opposite to the traveler's velocity with respect to the Earth's inertial coordinates.

However, this quantity does *not* equal the derivative of the proper distance with respect to the proper time. The proper distance s from the Earth in terms of the traveler's momentarily co-moving inertial coordinates at the proper time τ is

$$s(\tau) = \frac{\cosh(g\tau) - 1}{g}\sqrt{1 - \tanh(g\tau)^2} = \frac{\cosh(g\tau) - 1}{g\cosh(g\tau)}$$

which shows that the proper distance approaches a constant 1/g (about 1 light-year) as τ increases. This shouldn't be surprising, because we've already seen that the traveler's proper distance from a fixed point on the other side of the Earth actually is constant and equal to 1/g throughout the period of constant proper acceleration. The derivative of the proper distance of the Earth with respect to the proper time is

$$\frac{ds}{d\tau} = \frac{\tanh(g\tau)}{\cosh(g\tau)}$$

This can be regarded as a kind of velocity, since it represents the proper rate of change of the proper distance from the Earth as the traveller accelerates away. A plot of this function as τ varies from 0 to 6 years is shown below.

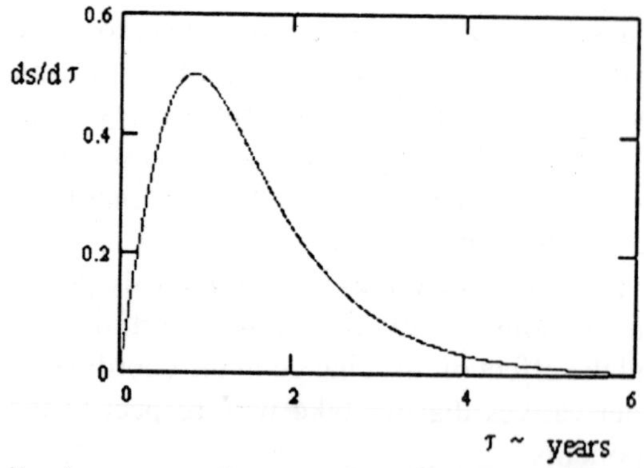

Initially the proper distance from the Earth increases as the traveller accelerates away, but eventually (if the constant proper acceleration is maintained for a sufficiently long time) the "length contraction" effect of his increasing velocity becomes great enough to cause the derivative to drop off to zero as the proper distance approaches a constant 1/g.

To find the point of maximum ds/dτ we differentiate again with respect to τ to give

$$\frac{d^2s}{d\tau^2} = -g\frac{\cosh(g\tau)^2 - 2}{\cosh(g\tau)^3}$$

Setting this to zero, we see that the maximum occurs at $\tau = \text{invcosh}\left(\sqrt{2}\right)/g$, and substituting this into the expression for ds/dτ gives the maximum value of 1/2. Thus the derivative of proper distance from Earth with respect to proper time during a constant 1g acceleration away from the Earth reaches a maximum of half the speed of light at a proper time of about 0.856 years, after which is drops to zero.

Similarly, the traveler's proper distance S from the turnaround point is given by

$$S(t) = \left[1\frac{\cosh(\text{inv}\sinh(gT)) - 1}{g} - \frac{\cosh(g\tau) - 1}{g}\right]\sqrt{1 - \tanh(g\tau)}^2$$

$$= \frac{1}{g}\left(\frac{2\sqrt{1 + g^2T^2} - 1}{\cosh(g\tau)} - 1\right).$$

The derivative of this with respect to the traveler's proper time is

$$\frac{dS}{d\tau} = -\tanh(g\tau)\frac{2\sqrt{1+g^2T^2}-1}{\cosh(g\tau)}$$

A plot of this "velocity" is shown below for the first quartile leg of a journey as described above with T = 20 years.

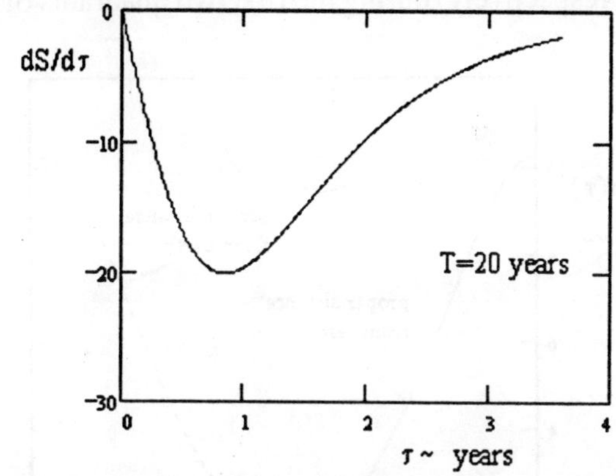

The magnitude of this "velocity" increases rapidly at the start of the acceleration, due to the combined effects of the traveler's motion and the onset of "length contraction", but if allowed to continue long enough the "velocity" drops off and approaches 2 (i.e., twice the speed of light) at the point where the traveller reverses his acceleration. Of course, the fact that this derivative exceeds c does not conflict with the fact that c is an upper limit on velocities with respect to inertial coordinate systems, because S and τ do not constitute inertial coordinates.

To find the extreme point on this curve we differentiate again with respect to τ, which gives

$$\frac{dS}{d\tau} = g\frac{\left(\cosh(g\tau)^2-2\right)\left(2\sqrt{1+g^2T^2}-1\right)}{\cosh(g\tau)^3}$$

Consequently we see that the extreme value occurs (assuming the journey is long enough and the acceleration is great enough) at the proper time $\tau = \text{invcosh}\,(\sqrt{2})/g$, where the value of dS/dτ is

$$\left(\frac{dS}{d\tau}\right)_{min} = \frac{1}{2} - \sqrt{1+g^2T^2}$$

By symmetry, these same two characteristics apply to all four of the "quadrants" of the traveler's journey, with the appropriate changes of sign and direction. The figure below shows the proper distances s(t) and S(t) (i.e., the distances from the origin and the destination respectively) during the first two quadrants of a journey with T = 6.

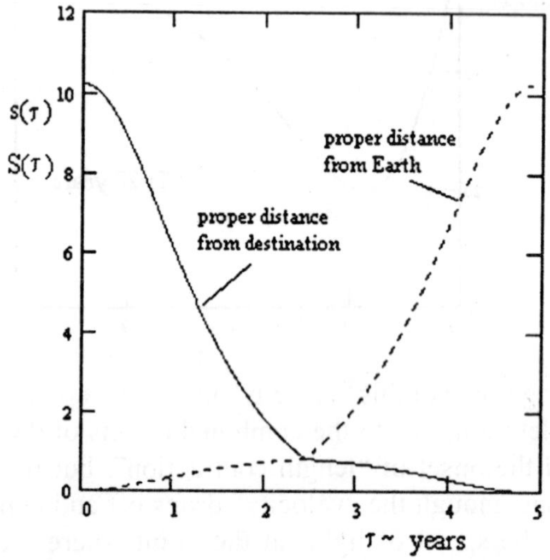

By symmetry we see that the portions of these curves to the right of the mid-point can be generated from the relation $s(\tau) = S(\tau_C - \tau)$. Also, it's obvious that

$$s(\tau) + S(\tau) = \sqrt{1 - v(\tau^2)}x_c$$

If we consider journeys with non-constant proper accelerations, it's possible to construct some slightly peculiar-sounding scenarios. For example, suppose the traveller accelerates in such a way that his velocity is 1 – exp(-kt) for some constant k. It follows that the distance in the Earth's frame at time t is [kt + exp(-kt) – 1]/k, so the distance in the traveler's frame is

$$\text{dis tan ce} = \sqrt{1-\left(1-e^{-kt}\right)^2}\,\frac{(kt + e^{-kt} - 1)}{k}$$

This function initially increases, then reaches a maximum, and then asymptotically approaches zero. With k = 1 year^{-1} the maximum occurs at roughly 3 years and a distance of about 0.65 light-years (relative to the traveler's frame). Thus we have the seemingly paradoxical situation that the Earth "becomes closer" to the traveller as he moves further away.

This is not as strange as it may sound at first. Suppose we leave home and drive for 1 hour at a constant speed of 20 mph. We could then say that we are "1 hour from home". Now suppose we suddenly accelerate to 40 mph.

How far (in time) are we away from home? If we extrapolate our current worldline back in time, we are only 1/2 hour from home. If we speed up some more, our "distance" (in terms of time) from home becomes less and less.

Of course, we have to speed up at a rate that more than compensates for the increasing road distance, but that's not hard to do (in theory).

The only difference between this scenario and the relativistic one is that when we accelerate to relativistic speeds both our time and our space axes are affected, so when we extrapolate our current frame of reference back to Earth we find that both the time and the distance are shortened.

Another interesting acceleration profile is the one that results from a constant nozzle velocity u and constant exhaust mass flow rate $w = dm_0/d\tau$, where τ is the proper time of the rocket, the effective force is uw throughout the acceleration. This does not result in constant proper acceleration, because the rest mass of the rocket is being reduced while the applied proper force remains constant. In this case we have

$$uw = \frac{m_0(\tau)}{(1-v(\tau)^2)^{3/2}}\frac{dv}{dt} = \frac{m_0(\tau)}{1-v(\tau)^2}\frac{dv}{d\tau}$$

where t is the time of the initial coordinates and v is the velocity of the rocket with respect to those coordinates. Also, we have $m_0(\tau) = m_0(0) - w\,\tau$, so we can integrate to get the speed

$$\int_0^{\tau_f} \frac{uw}{m_0(0)-w\tau}d\tau = \int_0^{v_f} \frac{1}{1-v^2}dv$$

Letting $\rho(\tau)$ denote the ratio $[m(0) - w\,\tau]/m(0)$, which is the ratio of rest masses at the start of the acceleration to the rest mass at proper time τ, the result is

$$\rho(\tau)^{2u} = \frac{1 - v(\tau)}{1 + v(\tau)}$$

so we have

$$v(\tau) = \frac{1 - \rho(\tau)^{2u}}{1 + \rho(\tau)^{2u}}$$

Also, since $dt = d\tau / \sqrt{1 - v^2}$, we can integrate this to get the coordinate time t as a function of the rocket's proper time

$$t(t) = \frac{m_0(0)}{2w\left(1 - u^2\right)}[2 - (1 + u)\rho(\tau)^{1-u} - (1 - u)\rho(\tau)^{1+u}]$$

$$= \frac{m_0(0)}{w\left(1 - u^2\right)}\left(1 - \rho\left[\cosh\left(u\ln(\rho)\right) - u\sinh(u\ln(\rho))\right]\right)$$

In the limit as the nozzle velocity u approaches 1, this expression reduces to

$$t = \frac{m_0(0)}{4w}\left[1 - 2\ln(\rho) - \rho^2\right].$$

It's interesting that for photonic propulsion (u=1) the mass ratio r is identical to the Doppler frequency shift of the exhaust photons relative to the original rest frame, i.e., we have

$$\rho(\tau) = \sqrt{\frac{1 - v(\tau)}{1 + v(\tau)}}.$$

Thus if the rocket continues to convert its own mass to energy and eject it as photons of a fixed frequency, the energy of each photon as seen from the fixed point of origin is exactly proportional to the rest mass of the rocket at the moment when the photon was ejected.

Also, since r(t) is the current rest mass $m_0(t)$ divided by the original rest mass $m_0(0)$, and since the inertial mass m(t) is related to the rest mass $m_0(t)$ by the equation $m(t) = m_0(t)/\sqrt{1 - v^2}$, we find that the inertial mass m(t) of the rocket is given as a function of the rocket's velocity v by the equation

$$m(v) = \frac{m_0(0)}{1+v}$$

Thus we find that as the rocket's velocity goes to 1 at the moment when it is converting the last of its rest mass into energy, so its rest mass is going to zero, its inertial mass goes to $m_0(0)/2$, i.e., exactly half of the rocket's original rest mass.

This is to be expected, because momentum must be conserved, and all the photons except that very last have been ejected in the rearward direction at the speed of light, leaving only the last remaining photon (which has nothing to react against) moving in the forward direction, so it must have momentum equal to all the rearward momentum of the ejected photons.

The momentum of a photon is $p = h\nu/c = E/c$, so in units with $c = 1$ we have $p = E$. The original energy content of the rocket was it's rest mass, $m_0(0)$, which has been entirely converted to energy, half in the forward direction (in the last remaining super-energetic photon) and half in the rearward direction (the progressively more redshifted stream of exhaust photons).

The preceding discussion focused on purely linear motion, but we can just as well consider arbitrary accelerated paths. It's trivial to determine the lapse of proper time along any given timelike path as a function of an inertial time coordinate simply by integrating $d\tau$ over the path, but it's a bit more challenging to express the lapse of proper time along one arbitrary worldline with respect to the lapse of proper time along another, because the appropriate correspondence is ambiguous.

Perhaps the most natural correspondence is given by mapping the proper time along the reference worldline to the proper time along the subject worldline by means of the instantaneously co-moving planes of inertial simultaneity of the reference worldline. In other words, to each point along the reference worldline we can assign a locus of simultaneous points based on co-moving inertial coordinates at that point, and we can then find the intersections of these loci with the subject worldline.

Quantitatively, suppose the reference worldline W_1 is given parametrically by the functions $x_1(t)$, $y_1(t)$, $z_1(t)$ where x, y, z, t are inertial coordinates. From this we can determine the derivatives $\dot{x}_1 = dx_1/dt$, $\dot{y}_1 = dy_1/dt$, and $\dot{z}_1 = dz_1/dt$. These also represent the components of the gradient of the space of simultaneity of the instantaneously co-moving inertial frame of the object. In other words, the spaces of simultaneity for W_1 have the partial derivatives

$$\frac{\partial t}{\partial x} = \dot{x}_1 \qquad \frac{\partial t}{\partial y} = \dot{y}_1 \qquad \frac{\partial t}{\partial z} = \dot{z}_1$$

These enable us to express the total differential time as a function of the differentials of the spatial coordinates

$$dt = \frac{\partial t}{\partial x}dx + \frac{\partial t}{\partial y}dy + \frac{\partial t}{\partial z}dz$$

If the subject worldline W_2 is expressed parametrically by the functions $x_2(t)$, $y_2(t)$, $z_2(t)$, and if the inertial plane of simultaneity of the event at coordinate time t_1 on W_1 is intersected by W_2 at the coordinate time t_2, then the difference in coordinate times between these two events can be expressed in terms of the differences in their spatial coordinates by substituting into the above total differential the quantities $dt = t_2 - t_1$, $dx = x_2(t_2) - x_1(t_1)$ and so on. The result is

$$t_2 - t_1 = \dot{x}_1\left[x_2(t_2) - x_1(t_1)\right] + \dot{y}_1\left[y_2(t_2) - y_1(t_1)\right] + \dot{z}_1 \left[z_2(t_2) - z_1(t_1)\right]$$

where the derivatives of x_1, y_1, and z_1 are evaluated at t_1. Rearranging terms and omitting the indications of functional dependence for the W_1 coordinates, this can be written in the form

$$t_2 - \dot{x}_1 x_2(t_2) - \dot{y}_1 y_2(t_2) - \dot{z}_1 z_2(t_2) = t_1 - \dot{x}_1 x_2 - \dot{y}_1 y_1 - \dot{z}_1 z_1$$

This is an implicit formula for the value of t_2 on W_2 corresponding to t_1 on W_1 based on the instantaneous inertial simultaneity of W_1. Every quantity in this equation is an explicit function of either t_1 or t_2, so we can solve for t_2 to give a function F_1 such that $t_2 = F_1(t_1)$. We can also integrate the absolute intervals along the two worldlines to give the functions f_1 and f_2 which relate the proper times along W_1 and W_2 to the coordinate time, i.e., we have $\tau_1 = f_1(t)$ and $\tau_2 = f_2(t)$. With these substitutions we arrive at the general form of the expression for τ_2 with respect to τ_1:

$$\tau_2 = f_2\left(F_1\left(f_1^{-1}(\tau_1)\right)\right)$$

To illustrate, suppose W_1 is the worldline of a particle moving along some arbitrary path and W_2 is just the worldline of the spatial origin of the inertial coordinates. In this case we have $x_2 = y_2 = z_2 = 0$ and $\tau_2 = t_2$, so the above formula reduces to

$$\tau_2 = t_1 - \dot{x}_1 x_1 - \dot{y}_1 y_1 - \dot{z}_1 z_1 = t_1 - r.v$$

where r and v are the position and velocity vectors of W_1 with respect to the inertial rest coordinates of W_2. Differentiating with respect to t_1, and multiplying through by $dt_1/d\tau_1 = (1-v^2)^{-1/2}$, we get

$$\frac{d\tau_2}{d\tau_1} = \frac{1 - v^2 - r.a}{\sqrt{1-v^2}} = \frac{1 - v^2 - ra\cos(\theta)}{\sqrt{1-v^2}}$$

where a is the acceleration vector and θ is the angle between the r and a vectors. Thus if the acceleration of W_1 is zero, we have $d\tau_2/d\tau_1 = (1-v^2)^{1/2}$. On the other hand, if W_2 is moving around W_1 in a circle at constant speed, we have $a = -v^2/r$ and the position and acceleration vectors are perpendicular, giving the result $d\tau_2/d\tau_1 = (1-v^2)^{-1/2}$.

This is consistent with the fact that, if the object is moving tangentially, the plane of simultaneity for its instantaneously co-moving inertial coordinate system intersects with the constant-t plane along the line from the object to the origin, and hence the time difference is entirely due to the transverse dilation (i.e., the square root of $1-v^2$ factor).

If the speed v of W_1 is constant, then we have the explicit equation

$$\tau_2 = \frac{\tau_1}{\sqrt{1-v^2}} - (x\dot{x} + y\dot{y} + z\dot{z})$$

To illustrate, suppose the object whose worldline is W_2 begins at the origin at t = 0 and thereafter moves counter-clockwise in a circle tangent to the origin in the xy plane with a constant angular velocity ω as illustrated below.

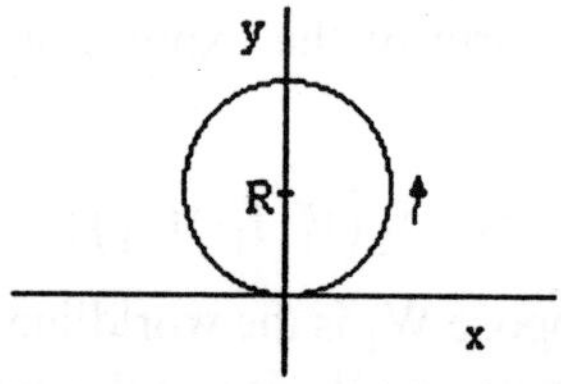

In this case the object's spatial coordinates and their derivatives as a function of coordinate time are

$$x_1 = R\sin(\omega t) \quad y_1 = R(1-\cos(\omega t))$$

$$\dot{x}_1 = R\omega\cos(\omega t) \quad \dot{y}_1 = R\omega\sin(\omega t)$$

Substituting into the equation for τ_2 and replacing each appearance of t with $\tau_1/\sqrt{1-v^2}$ gives the result

$$\tau_1 = \frac{\tau_1}{\sqrt{1-v^2}} - vR\sin\left(\frac{v\tau_1}{R\sqrt{1-v^2}}\right)$$

$$A^2\frac{\partial^2\phi}{\partial x^2} + 2AC\frac{\partial^2\phi}{\partial x\partial t^2} + C^2\frac{\partial^2\phi}{\partial t^2} = B^2\frac{\partial^2\phi}{\partial x^2} + 2BD\frac{\partial^2\phi}{\partial x\partial t} + D^2\frac{\partial^2\phi}{\partial t^2}$$

This is the proper time of the spatial origin according to the instantaneous time slices of the moving object's proper time. This function is plotted below with R = 1 and v = 0.8. Also shown is the stable component $\tau/\sqrt{1-v^2}$.

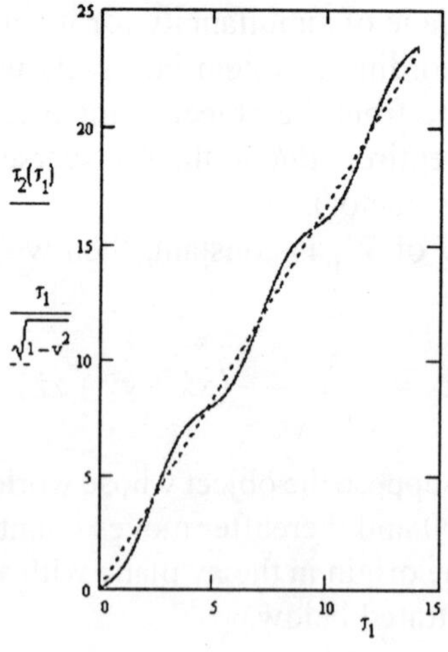

Naturally if the circle radius R goes to infinity the value of the sine function approaches the argument, and so the above expression reduces to

$$\tau_2 = \tau_1\sqrt{1-v^2}$$

This confirms the reciprocity between the two worldlines when both are inertial. We can also differentiate the full expression for τ_2 as a function of τ_1 to give the relation between the differentials

$$\frac{d\tau_2}{d\tau_1} = \frac{1 - v^2\cos\left(\dfrac{v\tau_1}{R\sqrt{1-v^2}}\right)}{\sqrt{1-v^2}}$$

This relation is plotted in the figure below, again for R = 1 and v = 0.8.

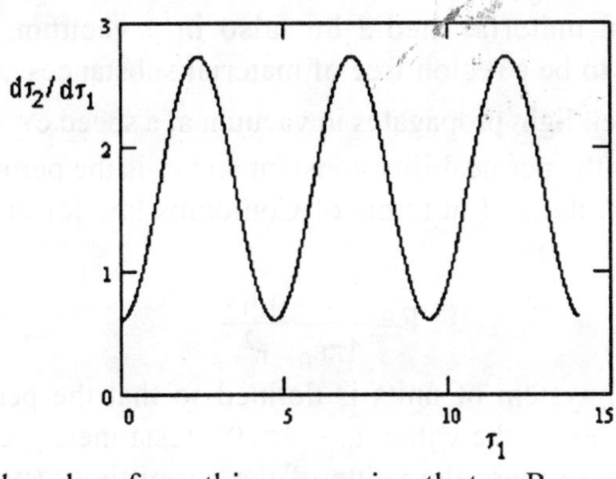

It's also clear from this expression that as R goes to infinity the cosine approaches 1, and we again have

$$\tau_2 = \tau_1\sqrt{1-v^2}$$

Incidentally, the above equation shows that the ratio of time rates equals 1 when the moving object is a circumferential distance of

$$\frac{v\tau_1}{\sqrt{1-v^2}} = \pm R\,\mathrm{a}\cos\left(\frac{1-\sqrt{1-v^2}}{v^2}\right) = \pm R\left(\frac{\pi}{3} - \left(\frac{\sqrt{3}}{12}\right)v^2 + \ldots\right)$$

from the point of tangency. Hence, for small velocities v the configuration of "equal time rates" occurs when the moving object is at $\pm\pi/3$ radians from the point of tangency. On the other hand, as v approaches 1, the configuration of equal time rates occurs when the moving object approaches the point of tangency.

This may seem surprising at first, because we might expect the proper time of the origin to be dilated with respect to the proper time of the tangentially moving object. However, the planes of simultaneity of the moving object are tilting very rapidly in this condition, and this offsets the usual time dilation factor. As v approaches 1, these two effects approach equal magnitude, and cancel out for a location approaching the point of tangency.

THE STARRY MESSENGER

Maxwell's equations are very successful at describing the propagation of light based on the model of electromagnetic waves, not only in material media but also in a vacuum, which is considered to be a region free of material substances. According to this model, light propagates in vacuum at a speed $c = 1/\sqrt{\varepsilon_0\mu_0}$, where μ_0 is the permeability constant and e_0 is the permittivity of the vacuum, defined in terms of Coulombs law for electrostatic force

$$F = \frac{1}{4\pi\varepsilon_0}\frac{q_1 q_2}{r^2}$$

The SI system of units is defined so that the permeability constant takes on the value $\mu_0 = 4\pi 10^{-7}$ tesla meter per ampere, and we can measure the value of the permittivity (typically by measuring the capacitance C between parallel plates of area A separated by a distance d, using the relation $\varepsilon_0 = Cd/A$) to have the value $\varepsilon_0 = (8.854187818)10^{-12}$ coulombs^2 per newton meters^2. This leads to the familiar value

$$c = \frac{1}{\sqrt{\varepsilon_0\mu_0}} = 299792458 \text{ meters/sec}$$

for the speed of light in a vacuum. Of course, if we place some substance between our capacitors when determining ε_0 we will generally get a different value, so the speed of light is different in various media. This leads to the index of refraction of various transparent media, defined as $n = c_{vacuum} / c_{medium}$. Thus Maxwell's theory of electro-magnetism seems to clearly imply that the speed of propagation of such electromagnetic waves depends only on the medium, and is independent of the speed of the source.

On the other hand, it also suggests that the speed of light depends on the motion of the medium, which is easy to imagine in the case of a material medium like glass, but not so easy if the "medium" is the vacuum of empty space. How can we even assign a state of motion to the vacuum? In struggling to answer this question, people tried to imagine that even the vacuum is permeated with some material-like substance, the ether, to which a definite state of motion could be assigned.

On this basis it was natural to suppose that Maxwell's equations were strictly applicable (and the speed of light was exactly c) *only* with respect to the absolute rest frame of the ether. With respect to other frames of reference they expected to find that the speed of light differed, depending on the direction of travel. Likewise we would expect to find corresponding differences and anisotropies in the capacitance of the vacuum when measured with plates moving at high speed relative to the ether.

However, when extremely precise interferometer measurements were carried out to find a directional variation in the speed of light on the Earth's surface (presumably moving through the ether at fairly high speed due to the Earth's rotation and its orbital motion around the Sun), essentially no directional variation in light speed was found that could be attributed to the motion of the apparatus through the ether. Of course, it had occurred to people that the ether might be "dragged along" by the Earth, so that objects on the Earth's surface are essentially at rest in the local ether.

However, these "convection" hypotheses are inconsistent with other observed phenomena, notably the aberration of starlight, which can only be explained in an ether theory if it is assumed that an observer on the Earth's surface is *not* at rest with respect to the local ether. Also, careful terrestrial measurements of the paths of light near rapidly moving massive objects showed no sign of any "convection". Considering all this, the situation was considered to be quite puzzling. There is a completely different approach that could be taken to modeling the phenomena of light, provided we're willing to reject Maxwell's theory of electromagnetic waves, and adopt instead a model similar to the one that Newton often seemed to have in mind, namely, an "emission theory".

One advocate of such a theory early in the early 1900's was Walter Ritz, who rejected Maxwell's equations on the grounds that the advanced potentials allowed by those equations were unrealistic.

Ritz debated this point with Albert Einstein, who argued that the observed asymmetry between advanced and retarded waves is essentially statistical in origin, due to the improbability of conditions needed to produce coherent advanced waves. Neither man persuaded the other. (Ironically, Einstein himself had already posited that Maxwell's equations were inadequate to fully represent the behaviour of light, and suggested a model that contains certain attributes of an emission theory to account for the photo-electric effect, but this challenge to Maxwell's equations was on a more subtle and profound level than Ritz's objection to advanced potentials).

In place of Maxwell's equations and the electromagnetic wave model of light, the advocates of "emission theories" generally assume a Galilean or Newtonian spacetime, and postulate that light is emitted and propagates away from the source (perhaps like Newtonian corpuscles) at a speed of c relative to the source. Thus, according to emission theories, if the source is moving directly toward or away from us with a speed v, then the light from that source is approaching us with a speed c+v or c–v respectively. Naturally this class of theories is compatible with experiments such as the one performed by Michelson and Morley, since the source of the light is moving along with the rest of the apparatus, so we wouldn't expect to find any directional variation in the speed of light in such experiments.

Also, an emission theory of light is compatible with stellar aberration, at least up to the limits of observational resolution. In fact, James Bradley (the discoverer of aberration) originally explained it on this very basis.

Of course, even an emission theory must account for the variations in light speed in different media, which means it can't simply say that the speed of light depends *only* on the speed of the source. It must also be dependent on the medium through which it is traveling, and presumably it must have a "terminal velocity" in each medium, i.e., a certain characteristic speed that it can maintain indefinitely as it propagates through the medium. (Obviously we never see light come to rest, nor even do we observe noticeable "slowing" of light in a given medium, so it must always exhibit a characteristic speed). Furthermore, based on the principles of an emission theory, the medium-dependent speed must be defined relative to the rest frame of the medium.

For example, if the characteristic speed of light in water is c_w, and a body of water is moving relative to us with a speed v, then (according to an emission theory) the light must move with a speed $c_w + v$ relative to us when it travels for some significant distance through that water, so that it has reached its "steady-state" speed in the water. In optics this distance is called the "extinction distance", and it is known to be proportional to $1/(\rho\lambda)$, where ρ is the density of the medium and λ is the wavelength of light. The extinction distance for most common media for optical light is extremely small, so essentially the light reaches its steady-state speed as soon as it enters the medium.

An experiment performed by Fizeau in 1851 to test for optical "convection" also sheds light on the viability of emission theories. Fizeau sent beams of light in both directions through a pipe of rapidly moving water to determine if the light was "dragged along" by the water. Since the refractive index of water is about $n = c/c_w = 1.33$ where c_w is the speed of light in water, we know that c_w equals c/1.33, which is about 75% of the speed of light in a vacuum. The question is, if the water is in motion relative to us, what is the speed (relative to us) of the light in the water?

If light propagates in an absolutely fixed background ether, and isn't dragged along by the water at all, we would expect the light speed to still be c_w relative to the fixed ether, regardless of how the water moves. This is admittedly a rather odd hypothesis (i.e., that light has a characteristic speed in water, but that this speed is relative to a fixed background ether, independent of the speed of the water), but it is one possibility that can't be ruled out *a priori*. In this case the difference in travel times for the two directions would be proportional to

$$\frac{1}{c_w} - \frac{1}{c_w} = 0$$

which implies no phase shift in the interferometer. On the other hand, if emission theories are right, the speed of the light in the water (which is moving at the speed v) should be c_w+v in the direction of the water's motion, and c_w-v in the opposite direction. On this basis the difference in travel times would be proportional to

$$\frac{1}{c_w - v} - \frac{1}{c_w + v} = \frac{2v}{c_w^2 - v^2}$$

This is a very small amount (remembering that c_w is about 75% of the speed of light in a vacuum), but it is large enough that it would be measurable with delicate interferometry techniques.

The results of Fizeau's experiment turned out to be consistent with *neither* of the above predictions. Instead, he found that the time difference (proportional to the phase shift) was a bit less than 43.5% of the prediction for an emission theory (i.e., 43.5% of the prediction based on the assumption of complete convection). By varying the density of the fluid we can vary the refractive index and therefore c_w, and we find that the measured phase shift always indicates a time difference of $(1-c_w^2)$ times the prediction of the emission theory. For water we have $c_w = 0.7518$, so the time lag is $(1-c_w^2) = 0.4346$ of the emission theory prediction.

This implies that if we let $S(c_w, v)$ and $S(c_w, -v)$ denote the speeds of light in the two directions, we have

$$\frac{1}{S(c_{w,} - v)} - \frac{1}{S(c_{w,} v)} = \frac{2v}{c_w^2 - v^2}\left(1 - c_w^2\right)$$

By partial fraction decomposition this can be written in the form

$$\frac{1}{S(c_{w,} - v)} - \frac{1}{S(c_{w,} v)} = \frac{A}{c_w - v} - \frac{B}{c_w + v}$$

where

$$(c_w + v)A - (c_w + v)B = 2v(1 - c_w^2)$$

Also, in view of the symmetry S(u, v) = S(v, u), we can swap c_w with v to give

$$(c_w + v)A + (c_w - v)B = 2c_w(1 - v^2)$$

Solving these last two equations for A and B gives A = 1 − vc_w and B = 1 + vc_w, so the function S is

$$S(u, v) = \frac{u + v}{1 + uv}$$

which of course is the relativistic formula for the composition of velocities. So, even if we rejected Maxwell's equations, it still appears that emission theories cannot be reconciled with Fizeau's experimental results.

More evidence ruling out simple emission theories comes from observations of a supernova made by Chinese astronomers in the year 1054 AD. When a star explodes as a supernova, the initial shock wave moves outward through the star's interior in just seconds, and elevates the temperature of the material to such a high level that fusion is initiated, and much of the lighter elements are fused into heavier elements, including some even heavier than iron. (This process yields most of the interesting elements that we find in the world around us). Material is flung out at high speeds in all directions, and this material emits enormous amounts of radiation over a wide range of frequencies, including x-rays and gamma rays.

Based on the broad range of spectral shifts (resulting from the Doppler effect), it's clear that the sources of this radiation have a range of speeds relative to the Earth of over 10000 km/sec. This is because we are receiving light emitted by some material that was flung out from the supernova in the direction away from the Earth, and by other material that was flung out in the direction toward the Earth.

If the supernova was located a distance D from us, then the time for the "light" (i.e., EM radiation of all frequencies) to reach us should be roughly D/c, where c is the speed of light. However, if we postulate that the actual speed of the light as it travels through interstellar space is affected by the speed of the source, and if the source was moving with a speed v relative to the Earth at the time of emission, then we would conclude that the light traveled at a speed of c+v on it's journey to the Earth.

Therefore, if the sources of light have velocities ranging from -v to +v, the first light from the initial explosion to reach the Earth would arrive at the time D/(c+v), whereas the last light from the initial explosion to reach the Earth would arrive at D/(c-v). This is illustrated in the figure below.

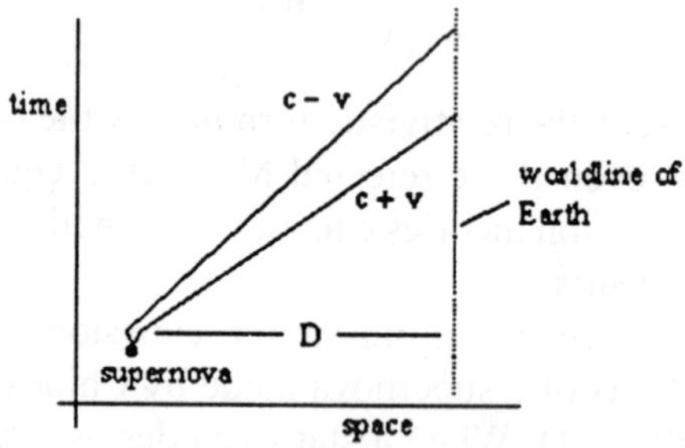

Hence the arrival times for light from the initial explosion event would be spread out over an interval of length D/(c−v) – D/(c+v), which equals $(D/c)(2v/c) / (1-(v/c)^2)$. The denominator is virtually 1, so we can say the interval of arrival times for the light from the explosion event of a supernova at a distance D is about (D/c)(2v/c), where v is the maximum speed at which radiating material is flung out from the supernova.

However, in actual observations of supernovae we do *not* see this "spreading out" of the event. For example, the Crab supernova was about 6000 light years away, so we had D/c = 6000 years, and with a range of source speeds of 10000 km/sec (meaning v = ±5000) we would expect a range of arrival times of 200 years, whereas in fact the Crab was only bright for less than a year, according to the observations recorded by Chinese astronomers in July of 1054 AD.

For a few weeks the "guest star", as they called it, in the constellation Taurus was the brightest star in the sky, and was even visible in the daytime for twenty-six days. Within two years it had disappeared completely to the naked eye. (It was not visible in Europe or the Islamic countries, since Taurus is below the horizon of the night sky in July for northern latitudes). In the time since the star went supernova the debris has expanded to it's present dimensions of about 3 light years, which implies that this material was moving at only (!) about 1/300 the speed of light.

Still, even with this value of v, the bright explosion event should have been visible on Earth for about 40 years (if the light really moved through space at $c \pm v$). Hence we can conclude that the light actually propagated through space at a speed essentially independent of the speed of the sources.

However, although this source independence of light speed is obviously consistent with Maxwell's equations and special relativity, we should be careful not to read too much into it. In particular, this isn't direct proof that the speed of light in a vacuum is independent of the speed of the source, because for visible light (which is all that was noted on Earth in July of 1054 AD) the extinction distance in the gas and dust of interstellar space is much less than the 6000 light year distance of the Crab nebula.

In other words, for visible light, interstellar space is not a vacuum, at least not over distances of many light years. Hence it's possible to argue that even if the initial speed of light in a vacuum was $c+v$, it would have slowed to c for most of its journey to Earth.

Admittedly, the details of such a counter-factual argument are lacking (because we don't really know the laws of propagation of light in a universe where the speed of light is dependent on the speed of the source, nor how the frequency and wavelength would be altered by interaction with a medium, so we don't know if the extinction distance is even relevant), but it's not totally implausible that the static interstellar dust might affect the propagation of light in such a way as to obscure the source dependence, and the extinction distance seems a reasonable way of quantifying this potential effect.

A better test of the source-independence of light speed based on astronomical observations is to use light from the high-energy end of the spectrum. As noted above, the extinction distance is proportional to $1/(\rho\lambda)$. For some frequencies of x-rays and gamma rays the extinction distance in interstellar space is about 60000 light years, much greater than the distances to many supernova events, as well as binary stars and other configurations with identifiable properties.

By observing these events and objects it has been found that the arrival times of light are essentially independent of frequency, e.g., the x-rays associated with a particular identifiable event arrive at the same time as the visible light for that event, even though the distance to the event is much less than the extinction distance for x-rays. This gives strong evidence that the speed of light in a vacuum is actually invariant and independent of the motion of the source.

With the aid of modern spectroscopy we can now examine supernovae events in detail, and it has been found that they exhibit several characteristic emission lines, particularly the signature of atomic hydrogen at 6563 angstroms. Using this as a marker we can determine the Doppler shift of the radiation, from which we can infer the speed of the source. The energy emitted by a star going supernova is comparable to all the energy that it emitted during millions or even billions of years of stable evolution.

Three main categories of supernovae have been identified, depending on the mass of the original star and how much of its "nuclear fuel" remains. In all cases the maximum luminosity occurs within just the first few days, and drops by 2 or 3 magnitudes within a month, and by 5 or 6 magnitudes within a year. Hence we can conclude that the light actually propagated through *empty* space at a speed essentially independent of the speed of the sources.

Another interesting observation involving the propagation of light was first proposed in 1913 by DeSitter. He wondered whether, if we assume the speed of light in a vacuum is always c with respect to the source, and if we assume a Galilean spacetime, we would notice anything different in the appearances of things. He considered the appearance of binary star systems, i.e., two stars that orbit around each other. More than half of all the visible stars in the night sky are actually double stars, i.e., two stars orbiting each other, and the elements of their orbits may be inferred from spectroscopic measurements of their radial speeds as seen from the Earth. DeSitter's basic idea was that if two stars are orbiting each other and we are observing them from the plane of their mutual orbit, the stars will be sometimes moving toward the Earth rapidly, and sometimes away. According to an emission theory this orbital component of velocity should be added to or subtracted from the speed of light. As a result, over the hundreds or thousands of years that it takes the light to reach the Earth, the arrival times of the light from approaching and receding sources would be very different.

Now, before we go any further, we should point out a potential difficulty for this kind of observation. The problem (again) is that the "vacuum" of empty space is not really a perfect vacuum, but contains small and sparse particles of dust and gas. Consequently it acts as a material and, as noted above, light will reach it's steady-state velocity with respect to that interstellar dust after having traveled beyond the extinction distance.

Since the extinction distance for visible light in interstellar space is quite short, the light will be moving at essentially c for almost its entire travel time, regardless of the original speed. For this reason, it's questionable whether visual observations of celestial objects can provide good tests of emission theory predictions. However, once again we can make use of the high-frequency end of the spectrum to strengthen the tests. If we focus on light in the frequency range of, say, x-rays and gamma rays, the extinction distance is much larger than the distances to many binary star systems, so we can carry out DeSitter's proposed observation (in principle) if we use x-rays, and this has actually been done by Brecher in 1977.

With the proviso that we will be focusing on light whose extinction distance is much greater than the distance from the binary star system to Earth (making the speed of the light simply c plus the speed of the star at the time of emission), how should we expect a binary star system to appear? Let's consider one of the stars in the binary system, and write its coordinates and their derivatives as

$$x(t) = D + R\cos(wt) \qquad \frac{dx}{dt} = -Rw\sin(wt)$$

$$y(t) = R\cos(wt) \qquad \frac{dx}{dt} = Rw\cos(wt)$$

where D is the distance from the Earth to the centre of the binary star system, R is the radius of the star's orbit about the system's centre, and w is the angular speed of the star. We also have the components of the emissive light speed

$$c^2 = c_x{}^2 + c_y{}^2$$

In these terms we can write the components of the absolute speed of the light emitted from the star at time t:

$$\dot{x} = -Rw\sin(wt) + c_x \qquad \dot{y} = Rw\cos(wt) + c_y$$

Now, in order to reach the Earth at time T the light emitted at time t must travel in the x direction from x(t) to 0 at a speed of for a time $\Delta t = T-t$, and similarly for the y direction. Hence we have

$$x(t) + \Delta t[-Rw\sin(wt) + c_x] = 0$$

$$y(t) + \Delta t[-Rw\sin(wt) + c_y] = 0$$

Substituting for x, y, and the light speed derivatives $\dot{x}$, $\dot{y}$, we have

$$D + R\cos(wt) - \Delta t\, Rw\sin(wt) = -\Delta t\, c_x$$

$$R\cos(wt) - \Delta t\, Rw\sin(wt) = -\Delta t\, c_y$$

Squaring both sides of both equations, and adding the resulting equations together, gives

$$D^2 + R^2 + (Dt\, Rw)^2 + 2DR[\cos(wt) - \Delta t\, w\sin(wt)] = \Delta t^2 c^2$$

Re-arranging terms gives the quadratic in Δt

$$\left[c^2 - R^2 w^2\right]\Delta t^2 + \left[2DRw\sin(wt)\right]$$

$$\Delta t - \left[D^2 + 2DR\cos(wt) + R^2\right] = 0$$

If we define the normalized parameters

$$v = \frac{Rw}{c}\, d = \frac{D}{c}\ r = \frac{R}{c}$$

then the quadratic in Δt becomes

$$\left[1 - v^2\right]\Delta t^2 + \left[2vd\sin(wt)\right]\Delta t - \left[d^2 + 2rd\cos(wt) + r^2\right] = 0$$

Solving this quadratic for $\Delta t = T-t$ and then adding t to both sides gives the arrival time T on Earth as a function of the emission time t on the star

$$T = t + \frac{-vd\sin(wt) + \sqrt{[vd\sin(wt)^2] + (1 - v^2)(d^2 + 2rd\cos(wt) + r^2)}}{1 - v^2}$$

If the star's speed v is much less than the speed of light, this can be expressed very nearly as

$$T = t - vd\sin(wt) + d + r\cos(wt)$$

The derivative of T with respect to t is

$$\frac{dT}{dt} = 1 - vwd\cos(wt) - rw\sin(wt)$$

and this takes it's minimum value when t = 0, where we have

$$\left(\frac{dT}{dt}\right)_{min} = 1 - vwd = 1 - \frac{d}{r}v^2$$

Consequently we find the DeSitter effect, i.e., dT/dt goes negative if $d > r / v^2$. Now, we know from Kepler's third law (which also applies in relativistic gravity with the appropriate choice of coordinates) that $m = r^3 w^2 = r v^2$, so we can substitute m/r for v^2 in our inequality to give the condition $d > r^2 / m$. Thus if the distance of the binary star system from Earth exceeds the square of the system's orbital radius divided by the system's mass (in geometric units) we would expect DeSitter's apparitions - assuming the speed of light is c ± v.

As an example, for a binary star system a distance of d = 20000 light-years away, with an orbital radius of r = 0.00001 light-years, and an orbital speed of v = 0.00005, the arrival time of the light as a function of the emission time is as shown below:

This corresponds to a star system with only about 1/6 solar mass, and an orbital radius of about 1.5 million kilometers. At any given reception time on Earth we can typically "see" at least three separate emission events from the same star at different points in its orbit. These ghostly apparitions are the effect that DeSitter tried to find in photographs of many binary star systems, but none exhibited this effect. He wrote

The observed velocities of spectroscipic doubles are as a matter of fact satisfactorily represented by a Keplerian motion.

Moreover in many cases the orbit derived from the radial velocities is confirmed by visual observations (as for d Equuli, z Herculis, etc). or by eclipse observations (as in Algol variables).

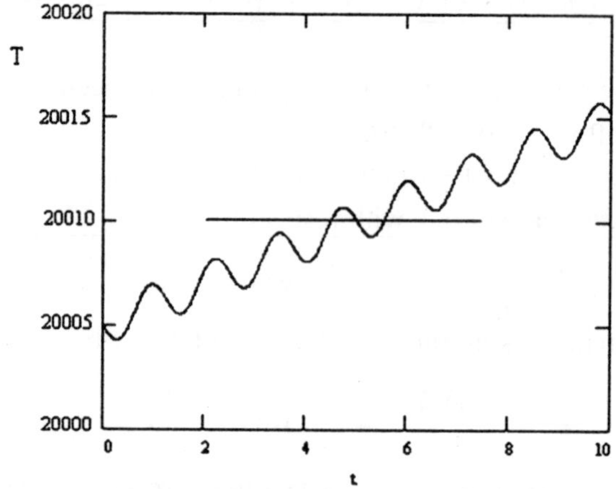

We can thus not avoid the conclusion [that] the velocity of light is independent of the motion of the source. Ritz's theory would force us to assume that the motion of the double stars is governed not by Newton's law, but by a much more complicated law, depending on the star's distance from the earth, which is evidently absurd.

Of course, he was looking in the frequency range of visible light, which we've noted is subject to extinction. However, in the x-ray range we can (in principle) perform the same basic test, and yet we still find no traces of these ghostly apparitions in binary stars, nor do we ever see the stellar components going in "reverse time" as we would according to the above profile. (Needless to say, for star systems at great distances it is not possible to distinguish the changes in transverse positions but, as noted above, by examining the Doppler shift of the radial components of their motions we can infer the motions of the individual bodies). Hence these observations support the proposition that the speed of light in empty space is essentially independent of the speed of the source.

In comparison, if we take the relativistic approach with constant light speed c, independent of the speed of the source, an analysis similar to the above gives the approximate result

$$T = t + d + r\cos(wt)$$

whose derivative is

$$\frac{dT}{dt} = 1 - v\sin(wt)$$

which is always positive for any v less than 1. This means we can't possibly have images arriving in reverse time, nor can we have any multiple appearances of the components of the binary star system.

Regarding this subject, Robert Shankland recalled Einstein telling him (in 1950) that he had himself considered an emission theory of light, similar to Ritz's theory, during the years before 1905, but he abandoned it because he could think of no form of differential equation which could have solutions representing waves whose velocity depended on the motion of the source. In this case the emission theory would lead to phase relations such that the propagated light would be all badly "mixed up" and might even "back up on itself". He asked me, "Do you understand that?" I said no, and he carefully repeated it all. When he came to the "mixed up" part, he waved his hands before his face and laughed, an open hearty laugh at the idea!

THOMAS PRECESSION

Consider a slanted rod AB in the xy plane moving at speed u in the positive y direction as indicated in the left-hand figure below. The A end of the rod crosses the x axis at time t = 0, whereas the B end does not cross until time t = 1. Hence we conclude that the rod is oriented at some non-zero angle with respect to the xyt coordinate system. However, suppose we view the same situation with respect to a system of inertial coordinates x'y't' (with x' parallel to x) moving in the positive x direction with speed v. In accord with special relativity, the x' and t' axes are skewed with respect to the x and t axes as shown in the right-hand figure below.

As a result of this skew, the B end of the rod crosses the x' axis at the same instant (i.e., the same t') as does the A end of the rod, which implies that the rod is parallel to the x' axis - and therefore to the x axis - based on the simultaneity of the x'y't' inertial frame. This implies that if a rod was parallel to the x axis and moving in the positive x direction with speed v, it would be perfectly aligned with the rod AB as the latter passed through the x' axis.

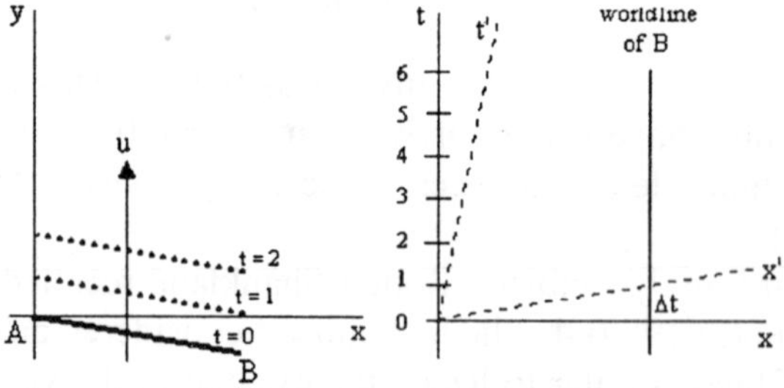

Thus if a rod is initially aligned with the x axis and moving with speed v in the positive x direction relative to a given fixed inertial frame, and then at some instant with respect to the rod's inertial rest frame it instantaneously changes course and begins to move purely in the positive y direction, without ever changing its orientation, we find that its orientation *does* change with respect to the original fixed frame of reference. This is because the changes in the states of motion of the individual parts of the rod do not occur simultaneously with respect to the original rest frame.

In general, whenever we transport a vector, always spatially parallel to itself in its own instantaneous rest frame, over an accelerated path, we find that its orientation changes relative to any given fixed inertial frame. This is the basic idea behind Thomas precession, named after Llewellyn Thomas, who first wrote about it in 1927.

For a simple application of this phenomenon, consider a particle moving around a circular path. The particle undergoes continuous acceleration, but at each instant it is at rest with respect to the momentarily co-moving inertial frame. If we consider the "parallel transport" of a vector around the continuous cycle of momentary inertial rest frames of the particle, we find that the vector does not remain fixed. Instead, it "precesses" as we follow it around the cycle.

This relativistic precession (which has no counter-part in non-relativistic physics) actually has observable consequences in the behaviour of sub-atomic particles.

To understand how the Thomas precession for simple circular motion can be deduced from the basic principles of special relativity, we can begin by supposing the circular path of a particle is approximated by an n-sided polygon, and consider the transition from one of these sides to the next, as illustrated below.

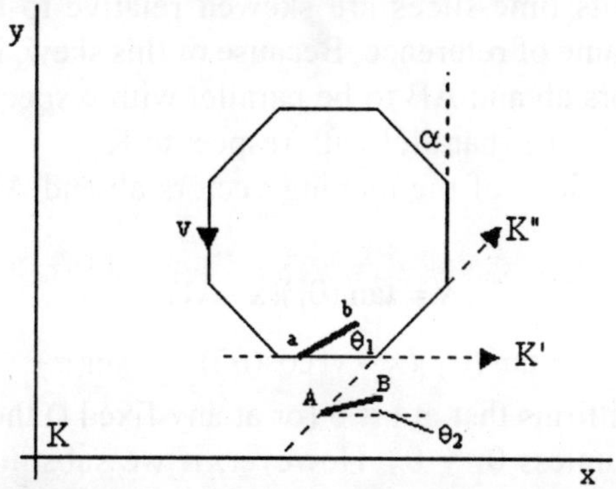

Let v denote the circumferential speed of the particle in the counter-clockwise direction, and note that $\alpha = 2\pi/n$ for an arbitrary n-sided regular polygon. (In the drawing above we have set n = 8). The dashed lines represent the loci of positions of the spatial origins of two inertial frames K' and K" that are co-moving with the particle on consecutive edges.

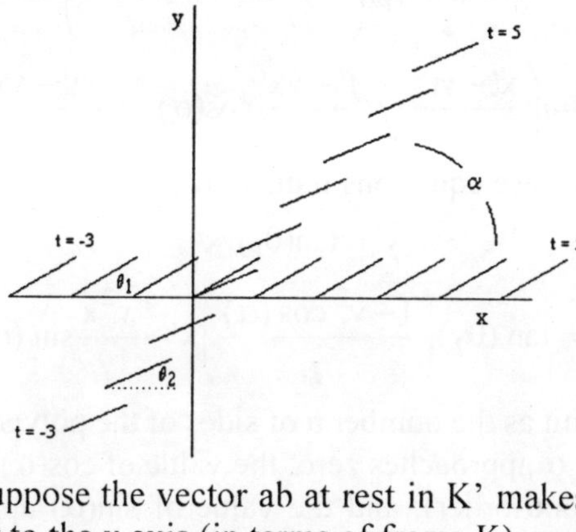

Now suppose the vector ab at rest in K' makes an angle θ_1 with respect to the x axis (in terms of frame K), and suppose the vector AB at rest in K" makes an angle of θ_2 with respect to the x axis. The figure below shows the positions of these two vectors at several consecutive instants of the frame K.

Clearly if θ_1 is not equal to θ_2, the two vectors will not coincide at the instant when their origins coincide. However, this assumes we use the definition of simultaneity associated with the inertial coordinate system K (i.e., the rest system of the polygon).

The system K' is moving in the positive x direction at the speed v, so its time-slices are skewed relative to those of the polygon's frame of reference. Because of this skew, it is possible for the vectors ab and AB to be parallel with respect to K' even though they are not parallel with respect to K.

The equations of the moving vectors ab and AB are easily seen to be

$$y = \tan(\theta_1)(x - vt)$$

$$y = \tan(\theta_2)(x - vt\cos(\alpha)) + vt\sin(\alpha)$$

This confirms that at t = 0 (or at any fixed t) these lines are not parallel unless $\theta_1 = \theta_2$. However, if we substitute from the Lorentz transformation between the frames K and K'

$$x = \frac{x' + vt'}{\gamma} \quad y = y' \quad t = \frac{t' + vx'}{\gamma}$$

where $\gamma = \sqrt{1 - v^2}$, the equations of the moving vectors become

$$y' = \tan(\theta_1)\left(\frac{x' + vt'}{\gamma} - v\frac{t' + vx'}{\gamma}\right)$$

$$y' = \tan(\theta_2)\left(\frac{x' + vt'}{\gamma} - v\frac{t' + vx'}{\gamma}\cos(\alpha)\right) + v\frac{t' + vx'}{\gamma}\sin(\alpha)$$

At t' = 0 these equations reduce to

$$y' = \tan(\theta_1)\gamma x'$$

$$y' = \tan(\theta_2)\left(\frac{1 - v^2\cos(\alpha)}{\gamma}\right)x' + \frac{v^2 x}{\gamma}\sin(\alpha)$$

In the limit as the number n of sides of the polygon increases and the angle α approaches zero, the value of $\cos(\alpha)$ approaches 1 (to the second order), and the value of $\sin(\alpha)$ approaches α. Hence the equations of the two moving vectors approach

$$y' = \tan(\theta_1)\gamma x' \quad y' = \tan(\theta_2)\gamma x' + \frac{v^2 x}{\gamma}\alpha$$

Setting these equal to each other, multiplying through by γ/x', and re-arranging, we get the condition

$$\tan(\theta_2) - \tan(\theta_1) = -\frac{v^2}{1-v^2}\alpha$$

Recalling the trigonometric identity

$$1 + \tan(\theta)^2 = \frac{1}{\cos(\theta)^2}$$

and noting that θ_1 approaches θ_2 in the limit as α goes to zero, the right-hand factor on the right side can be taken as

$$\tan(\theta_2) - \tan(\theta_1) = \tan(\theta_2 - \theta_1)[1 + \tan(\theta_1)\tan(\theta_2)]$$

where θ is the limiting value of both θ_1 and θ_2 as α goes to zero. Making use of these substitutions, and also noting that $\tan(\theta_2 - \theta_1)$ approaches $\theta_2 - \theta_1$, the condition for the two families of lines to be parallel with respect to frame K' (in the limit as α goes to zero) is

$$\theta_2 - \theta_1 = -\cos(\theta)^2 \frac{v^2}{1-v^2}\alpha$$

This is the amount by which the two vectors are skewed with respect to the K frame due to the transition around a single vertex of the polygon, given that the transported vector makes an angle θ with the edge leading into the vertex. The total precession resulting from one complete revolution around the n-sided polygon is n times the mean value of $\theta_2 - \theta_1$ for each of the n vertices of the polygon. Since $n = 2\pi/\alpha$, we can express the total precession as

$$\Delta\theta = -2\pi \frac{v^2}{1-v^2}\left\langle \cos(\theta)^2 \right\rangle_{mean}$$

If the circumferential speed v is small compared with 1, the denominator of this expression is closely approximated by 1, and the transported vector changes its absolute orientation only very slightly on one revolution.

In this case it follows that θ varies essentially uniformly from 0 to 2π as the vector is transported around the circle.

Hence for small v the total precession for one revolution is given closely by

$$\Delta\theta \approx -2\pi v^2 \left[\frac{1}{2\pi} \int_0^{2\pi} \cos(\theta)^2 \, d\theta \right] = -\pi v^2$$

On the other hand, if v is not small, we can consider the general situation illustrated below:

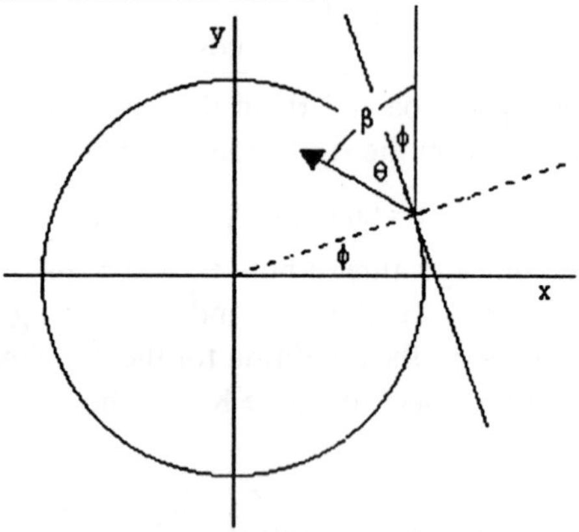

The variable ϕ signifies the absolute angular position of the transported vector at any given time, and β signifies the vector's orientation relative to the positive y axis. As before, θ denotes the angle of the vector relative to the local tangent "edge". We have the relations

$$\beta = \phi + \theta \qquad d\beta = d\phi + d\theta$$

We also have the following identifications involving the parameters ϕ and β:

$$d\phi = \alpha \qquad \frac{d\beta}{d\phi} = -\frac{v^2}{1-v^2}\cos(\theta)^2$$

Substituting $d\phi + d\theta$ for $d\beta$ and re-arranging, we get

$$\frac{-1}{1+\dfrac{v^2}{1-v^2}\cos(\theta)^2} d\theta = d\phi$$

This can be integrated explicitly to give ϕ as a function of θ.

Since β equals $\phi + \theta$, we can also give β as a function of θ, leading to the parametric equations

$$\phi(\theta) = -\gamma \,\text{inv}\tan(\gamma \tan(\theta)) \qquad \beta(\theta) = \theta + \phi(\theta)$$

where $\gamma = \sqrt{1-v^2}$. One complete "branch" is given by allowing θ to range from $-\pi/2$ to $\pi/2$, giving the angle ϕ from $-\gamma\pi/2$ to $\gamma\pi/2$, and the angles β from $-(\pi/2)(1-\gamma)$ to $(\pi/2)(1-\gamma)$. This is shown in the figure below.

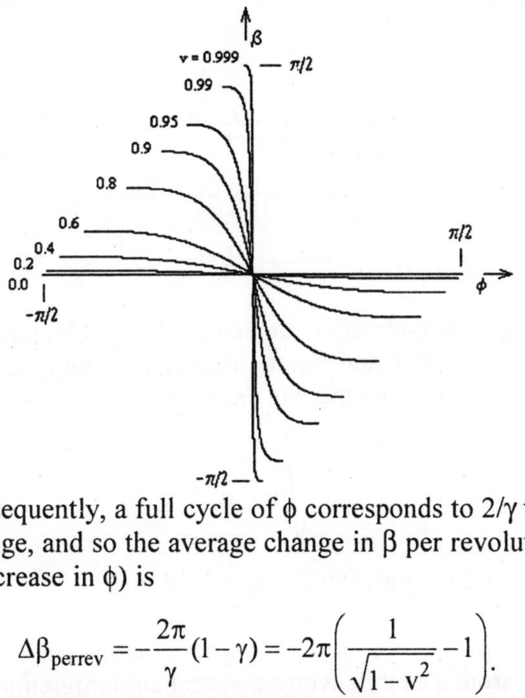

Consequently, a full cycle of ϕ corresponds to $2/\gamma$ times the above range, and so the average change in β per revolution (i.e., per 2π increase in ϕ) is

$$\Delta\beta_{\text{perrev}} = -\frac{2\pi}{\gamma}(1-\gamma) = -2\pi\left(\frac{1}{\sqrt{1-v^2}} - 1\right).$$

This function is plotted in the figure below, along with the "small v" approximation.

For all v less than 1 we can expand the general expression into a series

$$\Delta\beta \approx -\pi v^2\left(1 + \frac{3}{4}v^2 + \frac{5}{8}v^2 + \frac{35}{64}v^6 + \frac{63}{128}v^8 + \ldots\right)$$

These expressions represent the average change per revolution, because the cycles of β do not in general coincide with the cycles of ϕ. Resonance occurs when the ratio of the change in ϕ to the change in β is rational.

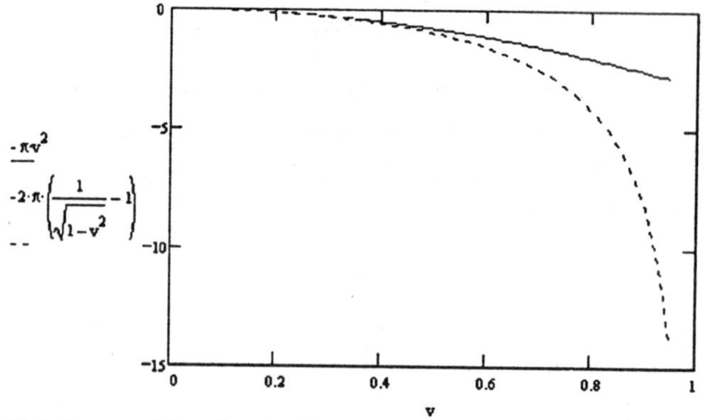

This is true if and only if there exist integers M, N such that

$$\frac{1}{\sqrt{1-v^2}} - 1 = \frac{M}{N}$$

Adding 1 to both sides, we can set 1 + (M/N) equal to m/n for integers m and n, and we can then square both sides and re-arrange to give, we find that the "resonant" values of v are given by

$$v = \sqrt{1 - \left(\frac{n}{m}\right)^2}$$

where m, n are integers with $|n|$ less than $|m|$.

We previously derived the low-speed approximation of the amount Thomas precession for a vector subjected to "parallel transport" around a circle with a constant circumferential speed v in the form $-\pi v^2$ radians per revolution. Dividing this by 2π gives the average precession rate of $-v^2/2$ in units of radians per radian (of travel around the circle).

We can also determine the average *rate* of Thomas precession, with units of radians per second. Letting ω_o denote the orbital angular velocity (i.e., the angular velocity with which the vector is transported around the circle of radius r), we have $v = \omega_o r$ and $a = v^2/r$ where a is the centripetal acceleration. Hence we have $\omega_o = v/r = a/v$, so multiplying $-v^2/2$ by ω_o gives the average Thomas precession rate $\omega_T = -va/2$ in units of rad/sec, which represents a frequency of $\nu_T = -(v^2/2)\nu_o = -va/(4\pi)$ cycles/sec.

Since the magnitude $-\pi v^2$ of the Thomas precession is of the second order in v, we might be tempted to think it is insignificant for ordinary terrestrial phenomena, but the expression $\nu_T = -(v^2/2)\nu_0$ shows that the precession frequency can be quite large in absolute terms, even if v is small, provided ν_0 is sufficiently large. This occurs when the orbital radius r is very small, giving a very large acceleration for any given orbital velocity. Consider, for example, the orbit of an electron around the nucleus of an atom. An electron has intrinsic quantum "spin" which tends to maintain it's absolute orientation much as does a spinning gyroscope, so it can be regarded as a vector undergoing parallel transport. Now, according to the original (naive) Bohr model, the classical orbit of an electron around the nucleus is given by equating the Coulomb and centripetal forces

$$ma = m\frac{v^2}{r} = \frac{e^2 N}{4\pi\varepsilon_0 r^2}$$

where e is the charge of an electron, m is the mass, e_0 is the permittivity of the vacuum, and N is the atomic number of the nucleus, so the linear and angular speeds of the electron are

$$v = \sqrt{\frac{e^2 N}{4\pi\varepsilon_0 r\, m}} \qquad \nu_0 = \frac{v}{2\pi r} = \sqrt{\frac{e^2 N}{16\pi^3\varepsilon_0 r^3\, m}}$$

Bohr hypothesized that the angular momentum L = mvr can only be an integer multiple of $h/(2\pi)$, so we have for some positive integer n

$$r = \left(\frac{nh}{e}\right)^2 \frac{\varepsilon_0}{\pi m N}.$$

Therefore, the linear velocity and orbital frequency of an electron (in this simplistic model) are

$$v = \frac{\alpha N}{n} \qquad \nu_0 = \frac{\alpha^2 N^2 m}{n^2 h}$$

where $\alpha = e^2/(2h\varepsilon_0)$ is the dimensionless "fine structure constant", whose value is approximately 1/137. (Remember that we are using units such that c = 1, so all distances are expressed in units of seconds). For the lowest energy state of a hydrogen atom we have n = N = 1, so the linear speed of the electron is about 1/137. Consequently the precession frequency is $-(v^2/2) = -0.00002664$ times the orbital frequency, which is a very small fraction, but it is still a very large frequency in absolute terms (1.755E-11 cycles/sec) because the orbital frequency is so large. (Note that these are *not* the frequencies of photons emitted from the atom, because those correspond to quanta of light given off due to transitions from one energy level to another, whereas these are the theoretical orbital frequencies of the electron itself in Bohr's simple model).

Incidentally, there is a magnetic interaction between the electron and nucleus of some atoms that is predicted to cause the electron's spin axis to precess by $+v^2$ radians per orbital radian, but the actual observed precession rate of the spin axes of electrons in such atoms is only $+(v^2/2)$.

For awhile after its discovery, there was no known explanation for this discrepancy. Only in 1927 did Thomas point out that special relativity implies the purely kinematic relativistic effect that now bears his name, which (as we've seen) yields a precession of $-(v^2/2)$ radians per orbital radian. The sum of this purely kinematic effect due to special relativity with the predicted effect due to the magnetic interaction yields the total observed $+(v^2/2)$ precession rate.

It's often said that the relativistic effect supplies a "factor of 2" (i.e., divides by 2) to the electron's precession rate. For example, Uhlenbeck wrote that

...when I first heard about [the Thomas precession], it seemed unbelievable that a relativistic effect could give a factor of 2 instead of something of order v/c... Even the cognoscenti of relativity theory (Einstein included!) were quite surprised.

(Uhlenbeck also told Pais that he didn't understand a word of Thomas's work when it first came out). However, this description is somewhat misleading, because (as we've seen) the relativistic effect is actually additive, not multiplicative. It just so happens that a particular magnetic interaction yields a precession of twice the frequency, and the opposite sign, as the Thomas precession, so the sum of the two effects is half the size of the magnetic effect alone. Both of the effects are second-order in the linear speed v/c. moving electrons (and the outside charge). Recall Gauss's Law relating the total electric field flux through a surface to the enclosed charge,

$$\int \vec{E}.d\vec{A} = (\text{enclosed charge})/\varepsilon_0.$$

Exercise: use this to verify that the electric field from an infinite line of charge λ coulombs per meter has magnitude $\lambda/2\pi r\varepsilon_0$ at distance r from the line charge.

From the exercise result, the electrostatic force on the charge q is:

$$F = qE = q\lambda\frac{v^2}{c^2}\frac{1}{2\pi r\varepsilon_0} = \frac{q\lambda v^2}{2\pi r}\frac{1}{\varepsilon_0 c^2} = \frac{q\lambda v^2}{2\pi r}\mu_0 \quad .$$

where in the last step we used $1/c^2 = \mu_0\varepsilon_0$.

This purely electrical force is identical in magnitude to the purely magnetic force in the other frame! So observers in the two frames *will* agree on the rate at which the particle accelerates away from the wire, but one will call the accelerating force magnetic, the other electric. We are forced to the conclusion that whether a particular force on an actual particle is magnetic or electric, or some mixture of both, depends on the frame of reference—so the distinction is rather artificial.

REMARKS ON GENERAL RELATIVITY

EINSTEIN'S PARABLE

In Einstein's little book *Relativity: the Special and the General Theory*, he introduces general relativity with a parable. He imagines going into deep space, far away from gravitational fields, where any body moving at steady speed in a straight line will continue in that state for a very long time. He imagines building a space station out there - in his words, "a spacious chest resembling a room with an observer inside who is equipped with apparatus." Einstein points out that there will be no gravity, the observer will tend to float around inside the room.

But now a rope is attached to a hook in the middle of the lid of this "chest" and an unspecified "being" pulls on the rope with a constant force. The chest and its contents, including the observer, accelerate "upwards" at a constant rate. How does all this look to the man in the room? He finds himself moving towards what is now the "floor" and needs to use his leg muscles to stand. If he releases anything, it accelerates towards the floor, and in fact all bodies accelerate at the same rate. If he were a normal human being, he would assume the room to be in a gravitational field, and might wonder why the room itself didn't fall. Just then he would discover the hook and rope, and conclude that the room was suspended by the rope.

Einstein asks: should we just smile at this misguided soul? His answer is no - the observer in the chest's point of view is just as valid as an outsider's. In other words, *being inside* the (from an outside perspective) *uniformly accelerating room is physically equivalent to being in a uniform gravitational field.* This is the basic postulate of general relativity. Special relativity said that all inertial frames were equivalent. General relativity extends this to accelerating frames, and states their equivalence to frames in which there is a gravitational field. This is called the *Equivalence Principle.*

The acceleration could also be used to cancel an existing gravitational field—for example, inside a freely falling elevator passengers are weightless, conditions are equivalent to those in the unaccelerated space station in outer space.

It is important to realise that this equivalence between a gravitational field and acceleration is only possible because the gravitational mass is exactly equal to the inertial mass. There is no way to cancel out electric fields, for example, by going to an accelerated frame, since many different charge to mass ratios are possible.

As physics has developed, the concept of fields has been very valuable in understanding how bodies interact with each other. We visualize the electric field lines coming out from a charge, and know that something is there in the space around the charge which exerts a force on another charge coming into the neighborhood.

We can even compute the energy density stored in the electric field, locally proportional to the square of the electric field intensity. It is tempting to think that the gravitational field is quite similar—after all, it's another inverse square field. Evidently, though, this is not the case. If by going to an accelerated frame the gravitational field can be made to vanish, at least locally, it cannot be that it stores energy in a simply defined local way like the electric field.

We should emphasize that going to an accelerating frame can only cancel a *constant* gravitational field, of course, so there is no accelerating frame in which the whole gravitational field of, say, a massive body is zero, since the field necessarily points in different directions in different regions of the space surrounding the body.

SOME CONSEQUENCES OF THE EQUIVALENCE PRINCIPLE

Consider a freely falling elevator near the surface of the earth, and suppose a laser fixed in one wall of the elevator sends a pulse of light horizontally across to the corresponding point on the opposite wall of the elevator. Inside the elevator, where there are no fields present, the environment is that of an inertial frame, and the light will certainly be observed to proceed directly across the elevator. Imagine now that the elevator has windows, and an outsider at rest relative to the earth observes the light. As the light crosses the elevator, the elevator is of course accelerating downwards at g, so since the flash of light will hit the opposite elevator wall at precisely the height relative to the elevator at which it began, the outside observer will conclude that the flash of light also accelerates downwards at g.

In fact, the light could have been emitted at the instant the elevator was released from rest, so we must conclude that light falls in an initially parabolic path in a constant gravitational field. Of course, the light is traveling very fast, so the curvature of the path is small! Nevertheless, *the Equivalence Principle forces us to the conclusion that the path of a light beam is bent by a gravitational field.*

The curvature of the path of light in a gravitational field was first detected in 1919, by observing stars very near to the sun during a solar eclipse. The deflection for stars observed very close to the sun was 1.7 seconds of arc, which meant measuring image positions on a photograph to an accuracy of hundredths of a millimeter, quite an achievement at the time.

One might conclude from the brief discussion above that a light beam in a gravitational field follows the same path a Newtonian particle would if it moved at the speed of light. This is true in the limit of small deviations from a straight line in a constant field, but is not true even for small deviations for a spatially varying field, such as the field near the sun the starlight travels through in the eclipse experiment mentioned above.

We could try to construct the path by having the light pass through a series of freely falling (fireproof!) elevators, all falling towards the centre of the sun, but then the elevators are accelerating relative to each other (since they are all falling along *radii*), and matching up the path of the light beam through the series is tricky. If it is done correctly (as Einstein did) it turns out that the angle the light beam is bent through is twice that predicted by a naove Newtonian theory.

What happens if we shine the pulse of light vertically *down* inside a freely falling elevator, from a laser in the centre of the ceiling to a point in the centre of the floor? Let us suppose the flash of light leaves the ceiling at the instant the elevator is released into free fall. If the elevator has height h, it takes time h/c to reach the floor. This means the floor is moving downwards at speed gh/c when the light hits.

Question: Will an observer on the floor of the elevator see the light as Doppler shifted?

The answer has to be no, because inside the elevator, by the Equivalence Principle, conditions are identical to those in an inertial frame with no fields present. There is nothing to change the frequency of the light. This implies, however, that to an outside observer, stationary in the earth's gravitational field, the frequency of the light *will* change.

This is because he will agree with the elevator observer on what was the initial frequency f of the light as it left the laser in the ceiling (the elevator was at rest relative to the earth at that moment) so if the elevator operator maintains the light had the same frequency f as it hit the elevator floor, which is moving at *gh/c* relative to the earth at that instant, the earth observer will say the light has frequency $f(1 + v/c) = f(1+gh/c^2)$, using the Doppler formula for very low speeds.

We conclude from this that light shining downwards in a gravitational field is shifted to a higher frequency. Putting the laser in the elevator floor, it is clear that light shining upwards in a gravitational field is red-shifted to lower frequency. Einstein suggested that this prediction could be checked by looking at characteristic spectral lines of atoms near the surfaces of very dense stars, which should be red-shifted compared with the same atoms observed on earth, and this was confirmed.

This has since been observed much more accurately. An amusing consequence, since the atomic oscillations which emit the radiation are after all just accurate clocks, is that *time passes at different rates at different altitudes*. The US atomic standard clock, kept at 5400 feet in Boulder, gains 5 microseconds per year over an identical clock almost at sea level in the Royal Observatory at Greenwich, England. Both clocks are accurate to one microsecond per year. This means you would age more slowly if you lived on the surface of a planet with a large gravitational field. Of course, it might not be very comfortable.

GENERAL RELATIVITY AND THE GLOBAL POSITIONING SYSTEM

Despite what you might suspect, the fact that time passes at different rates at different altitudes has significant practical consequences. An important *everyday* application of general relativity is the Global Positioning System. A GPS unit finds out where it is by detecting signals sent from orbiting satellites at precisely timed intervals. If all the satellites emit signals simultaneously, and the GPS unit detects signals from four different satellites, there will be three relative time delays between the signals it detects.

The signals themselves are encoded to give the GPS unit the precise position of the satellite they came from at the time of transmission. With this information, the GPS unit can use the speed of light to translate the detected time delays into distances, and therefore compute its own position on earth by triangulation.

But this has to be done very precisely! Bearing in mind that the speed of light is about one foot per nanosecond, an error of 100 nanoseconds or so could, for example, put an airplane off the runway in a blind landing. This means the clocks in the satellites timing when the signals are sent out must certainly be accurate to 100 nanoseconds a day. That is one part in 10^{12}.

It is easy to check that both the special relativistic time dilation correction from the speed of the satellite, and the general relativistic gravitational potential correction are much greater than that, so the clocks in the satellites must be corrected appropriately. (The satellites go around the earth once every twelve hours, which puts them at a distance of about four earth radii. The calculations of time dilation from the speed of the satellite, and the clock rate change from the gravitational potential, are left as exercises for the student

In fact, Ashby reports that when the first Cesium clock was put in orbit in 1977, those involved were sufficiently skeptical of general relativity that the clock was not corrected for the gravitational redshift effect.

But—just in case Einstein turned out to be right—the satellite was equipped with a synthesizer that could be switched on if necessary to add the appropriate relativistic corrections. After letting the clock run for three weeks with the synthesizer turned off, it was found to differ from an identical clock at ground level by precisely the amount predicted by special plus general relativity, limited only by the accuracy of the clock.

This simple experiment verified the predicted gravitational redshift to about one percent accuracy! The synthesizer was turned on and left on.

Chapter 6

Special Relativity

FRAMES OF REFERENCE AND NEWTON'S LAWS

The cornerstone of the theory of special relativity is the *Principle of Relativity*:

The Laws of Physics are the same in all inertial frames of reference.

We shall see that many surprising consequences follow from this innocuous looking statement.

Let us first, however, briefly review Newton's mechanics in terms of frames of reference.

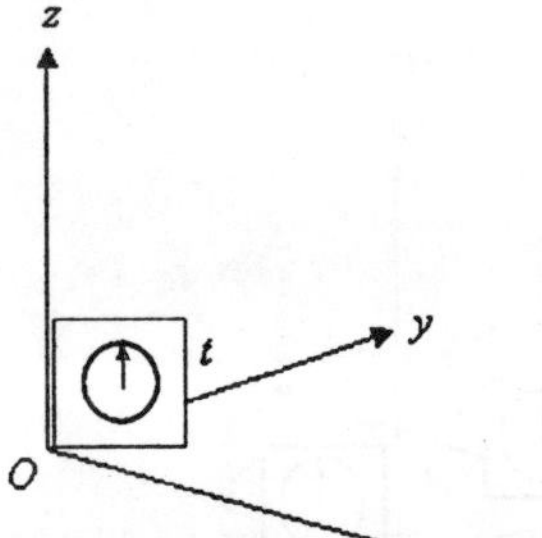

A frame of reference

A "frame of reference" is just a set of coordinates: something you use to measure the things that matter in Newtonian problems, that is to say, positions and velocities, so we also need a clock.

A point in space is specified by its three coordinates (x, y, z) and an "event" like, say, a little explosion, by a place and time: (x, y, z, t).

An inertial frame is defined as one in which Newton's law of inertia holds—that is, any body which isn't being acted on by an outside force stays at rest if it is initially at rest, or continues to move at a constant velocity if that's what it was doing to begin with. An example of a *non*-inertial frame is a rotating frame, such as a carousel.

The "laws of physics" we shall consider first are those of Newtonian mechanics, as expressed by Newton's Laws of Motion, with gravitational forces and also contact forces from objects pushing against each other. For example, knowing the universal gravitational constant from experiment (and the masses involved), it is possible from Newton's Second Law,

force = mass X acceleration,

to predict future planetary motions with great accuracy.

Suppose we know from experiment that these laws of mechanics are true in one frame of reference. How do they look in another frame, moving with respect to the first frame? To find out, we have to figure out how to get from position, velocity and acceleration in one frame to the corresponding quantities in the second frame.

Obviously, the two frames must have a constant relative velocity, otherwise the law of inertia won't hold in both of them. Let's choose the coordinates so that this velocity is along the x-axis of both of them.

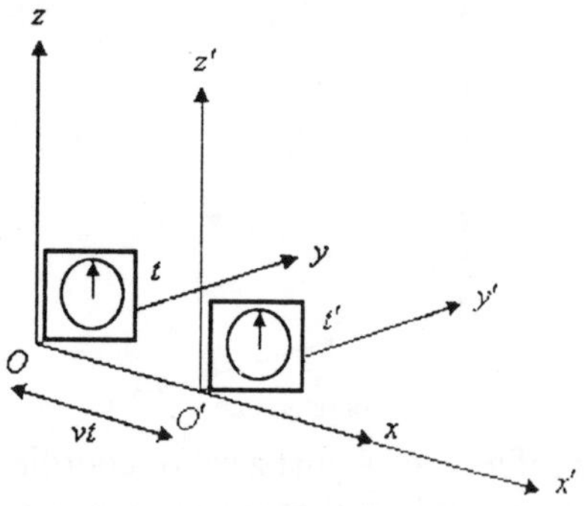

Fig. Two Frames of Reference Relatively Dispalced Along the x-axis

Notice we also throw in a clock with each frame. Suppose S' is proceeding relative to S at speed v along the x-axis. For convenience, let us label the moment when O' passes O as the zero point of timekeeping.

Now what are the coordinates of the event (x, y, z, t) in S' ? It's easy to see $t' = t$—we synchronized the clocks when O' passed

O. Also, evidently, $y' = y$ and $z' = z$, from the figure. We can also see that $x = x' + vt$. Thus (x, y, z, t) in S corresponds to (x', y', z', t') in S', where

$$x' = x - vt$$
$$y' = y$$
$$z' = z$$
$$t' = t$$

That's how *positions* transform; these are known as the *Galilean* transformations.

What about *velocities*? The velocity in S' in the x' direction

$$u_x' = \frac{dx'}{dt'} = \frac{dx'}{dt} = \frac{d}{dt}(x - vt) = \frac{dx}{dt} - v = u_x - v.$$

This is obvious anyway: it's just the addition of velocities formula

$$u_x = u_x' + v.$$

How does *acceleration* transform?

$$\frac{du_x'}{dt'} = \frac{du_x'}{dt} = \frac{d}{dt}(u_x - v) = \frac{du_x}{dt}$$

since v is constant.

That is to say,

$$a_x' = a_x$$

the *acceleration is the same* in both frames. This again is obvious—the acceleration is the rate of change of velocity, and the velocities of the same particle measured in the two frames differ by a *constant* factor-the relative velocity of the two frames.

If we now look at the motion under gravitational forces, for example,

$$m_1\vec{a} = \frac{Gm_1m_2}{r^2}\hat{r}$$

we get the same law on going to another inertial frame because *every term in the above equation stays the same*.

Note that $m\vec{a}$ is the rate of change of momentum—this is the same in both frames. So, in a collision, say, if total momentum is conserved in one frame (the sum of individual rates of change of momentum is zero) the same is true in *all* inertial frames.

THE SPEED OF LIGHT

EARLY IDEAS ABOUT LIGHT PROPAGATION

Attempts to measure the speed of light played an important part in the development of the theory of special relativity, and, indeed, the speed of light is central to the theory.

The first recorded discussion of the speed of light is in Aristotle, where he quotes Empedocles as saying the light from the sun must take some time to reach the earth, but Aristotle himself apparently disagrees, and even Descartes thought that light traveled instantaneously. Galileo, unfairly as usual, in *Two New Sciences* has Simplicio stating the Aristotelian position,

SIMP. Everyday experience shows that the propagation of light is instantaneous; for when we see a piece of artillery fired at great distance, the flash reaches our eyes without lapse of time; but the sound reaches the ear only after a noticeable interval.

Of course, Galileo points out that in fact nothing about the speed of light can be deduced from this observation, except that light moves faster than sound. He then goes on to suggest a possible way to measure the speed of light. The idea is to have two people far away from each other, with covered lanterns. One uncovers his lantern, then the other immediately uncovers his on seeing the light from the first.

This routine is to be practiced with the two close together, so they will get used to the reaction times involved, then they are to do it two or three miles apart, or even further using telescopes, to see if the time interval is perceptibly lengthened.

Galileo claims he actually tried the experiment at distances less than a mile, and couldn't detect a time lag. From this one can certainly deduce that light travels at least ten times faster than sound.

MEASURING THE SPEED OF LIGHT WITH JUPITER'S MOONS

The first real measurement of the speed of light came about half a century later, in 1676, by a Danish astronomer, Ole Rφmer, working at the Paris Observatory. He had made a systematic study of Io, one of the moons of Jupiter, which was eclipsed by Jupiter at regular intervals, as Io went around Jupiter in a circular orbit at a steady rate.

Actually, Rφmer found, for several months the eclipses lagged more and more behind the expected time, until they were running about eight minutes late, then they began to pick up again, and in fact after about six months were running eight minutes early. The cycle then repeated itself. Rφmer realized the significance of the time involved-just over one year.

This time period had nothing to do with Io, but was the time between successive closest approaches of earth in its orbit to Jupiter. The eclipses were furthest behind the predicted times when the earth was furthest from Jupiter.

The natural explanation was that the light from Io (actually reflected sunlight, of course) took time to reach the earth, and took the longest time when the earth was furthest away. From his observations, Rφmer concluded that light took about twenty-two minutes to cross the earth's orbit.

"For it is now certain from the phenomena of Jupiter's satellites, confirmed by the observations of different astronomers, that light is propagated in succession and requires about seven or eight minutes to travel from the sun to the earth." This is essentially the correct value.

Of course, to find the speed of light it was also necessary to know the distance from the earth to the sun. During the 1670's, attempts were made to measure the parallax of Mars, that is, how far it shifted against the background of distant stars when viewed simultaneously from two different places on earth at the same time.

This (very slight) shift could be used to find the distance of Mars from earth, and hence the distance to the sun, since all *relative* distances in the solar system had been established by observation and geometrical analysis. According to Crowe, they concluded that the distance to the sun was between 40 and 90 million miles.

Measurements presumably converged on the correct value of about 93 million miles soon after that, because it appears Romer (or perhaps Huygens, using Rφmer's data a short time later) used the correct value for the distance, since the speed of light was calculated to be 125, 000 miles per second, about three-quarters of the correct value of 186, 300 miles per second. This error is fully accounted for by taking the time light needs to cross the earth's orbit to be twenty-two minutes instead of the correct value of sixteen minutes.

Starlight and Rain

The next substantial improvement in measuring the speed of light took place in 1728, in England. An astronomer James Bradley, sailing on the Thames with some friends, noticed that the little pennant on top of the mast changed position each time the boat put about, even though the wind was steady.

He thought of the boat as the earth in orbit, the wind as starlight coming from some distant star, and reasoned that the apparent direction the starlight was "blowing" in would depend on the way the earth was moving. Another possible analogy is to imagine the starlight as a steady downpour of rain on a windless day, and to think of yourself as walking around a circular path at a steady pace.

The apparent direction of the incoming rain will not be vertically downwards-more will hit your front than your back. In fact, if the rain is falling at, say, 15 mph, and you are walking at 3 mph, to you as observer the rain will be coming down at a slant so that it has a vertical speed of 15 mph, and a horizontal speed towards you of 3 mph.

Whether it is slanting down from the north or east or whatever at any given time depends on where you are on the circular path at that moment. Bradley reasoned that the apparent direction of incoming starlight must vary in just this way, but the angular change would be a lot less dramatic. The earth's speed in orbit is about 18 miles per second, he knew from Rφmer's work that light went at about 10, 000 times that speed.

That meant that the angular variation in apparent incoming direction of starlight was about the magnitude of the small angle in a right-angled triangle with one side 10, 000 times longer than the other, about one two-hundredth of a degree. Notice this would have been just at the limits of Tycho's measurements, but the advent of the telescope, and general improvements in engineering, meant this small angle was quite accurately measurable by

Bradley's time, and he found the velocity of light to be 185, 000 miles per second, with an accuracy of about one percent.

Fast Flickering Lanterns

The problem is, all these astronomical techniques do not have the appeal of Galileo's idea of two guys with lanterns. It would be reassuring to measure the speed of a beam of light between two points on the ground, rather than making somewhat indirect deductions based on apparent slight variations in the positions of stars.

We can see, though, that if the two lanterns are ten miles apart, the time lag is of order one-ten thousandth of a second, and it is difficult to see how to arrange that. This technical problem was solved in France about 1850 by two rivals, Fizeau and Foucault, using slightly different techniques. In Fizeau's apparatus, a beam of light shone between the teeth of a rapidly rotating toothed wheel, so the "lantern" was constantly being covered and uncovered. Instead of a second lantern far away, Fizeau simply had a mirror, reflecting the beam back, where it passed a second time between the teeth of the wheel.

The idea was, the blip of light that went out through one gap between teeth would only make it back through the same gap if the teeth had not had time to move over significantly during the round trip time to the far away mirror. It was not difficult to make a wheel with a hundred teeth, and to rotate it hundreds of times a second, so the time for a tooth to move over could be arranged to be a fraction of one ten thousandth of a second.

The method worked. Foucault's method was based on the same general idea, but instead of a toothed wheel, he shone the beam on to a rotating mirror. At one point in the mirror's rotation, the reflected beam fell on a distant mirror, which reflected it right back to the rotating mirror, which meanwhile had turned through a small angle. After this second reflection from the rotating mirror, the position of the beam was carefully measured.

This made it possible to figure out how far the mirror had turned during the time it took the light to make the round trip to the distant mirror, and since the rate of rotation of the mirror was known, the speed of light could be figured out. These techniques gave the speed of light with an accuracy of about 1, 000 miles per second.

Albert Abraham Michelson

Albert Michelson was born in 1852 in Strzelno, Poland. His father Samuel was a Jewish merchant, not a very safe thing to be at the time. Purges of Jews were frequent in the neighboring towns and villages. They decided to leave town. Albert's fourth birthday was celebrated in Murphy's Camp, Calaveras County, about fifty miles south east of Sacramento, a place where five million dollars worth of gold dust was taken from one four acre lot. Samuel prospered selling supplies to the miners. When the gold ran out, the Michelsons moved to Virginia City, Nevada, on the Comstock lode, a silver mining town. Albert went to high school in San Francisco. In 1869, his father spotted an announcement in the local paper that Congressman Fitch would be appointing a candidate to the Naval Academy in Annapolis, and inviting applications. Albert applied but did not get the appointment, which went instead to the son of a civil war veteran.

However, Albert knew that President Grant would also be appointing ten candidates himself, so he went east on the just opened continental railroad to try his luck. Unknown to Michelson, Congressman Fitch wrote directly to Grant on his behalf, saying this would really help get the Nevada Jews into the Republican party. This argument proved persuasive. In fact, by the time Michelson met with Grant, all ten scholarships had been awarded, but the President somehow came up with another one.

Of the incoming class of ninety-two, four years later twenty-nine graduated. Michelson placed first in optics, but twenty-fifth in seamanship. The Superintendent of the Academy, Rear Admiral Worden, who had commanded the Monitor in its victory over the Merrimac, told Michelson: "If in the future you'd give less attention to those scientific things and more to your naval gunnery, there might come a time when you would know enough to be of some service to your country."

Michelson Measures the Speed of Light

On returning to Annapolis from the cruise, Michelson was

commissioned Ensign, and in 1875 became an instructor in physics and chemistry at the Naval Academy, under Lieutenant Commander William Sampson. Michelson met Mrs. Sampson's niece, Margaret Heminway, daughter of a very successful Wall Street tycoon, who had built himself a granite castle in New Rochelle, NY. Michelson married Margaret in an Episcopal service in New Rochelle in 1877.

At work, lecture demonstrations had just been introduced at Annapolis. Sampson suggested that it would be a good demonstration to measure the speed of light by Foucault's method. Michelson soon realized, on putting together the apparatus, that he could redesign it for much greater accuracy, but that would need money well beyond that available in the teaching demonstration budget.

He went and talked with his father in law, who agreed to put up $2, 000. Instead of Foucault's 60 feet to the far mirror, Michelson had about 2, 000 feet along the bank of the Severn, a distance he measured to one tenth of an inch. He invested in very high quality lenses and mirrors to focus and reflect the beam. His final result was 186, 355 miles per second, with possible error of 30 miles per second or so.

This was twenty times more accurate than Foucault, made the New York Times, and Michelson was famous while still in his twenties. In fact, this was accepted as the most accurate measurement of the speed of light for the next forty years, at which point Michelson measured it again.

THE MICHELSON-MORLEY EXPERIMENT

THE NATURE OF LIGHT

As a result of Michelson's efforts in 1879, the speed of light was known to be 186, 350 miles per second with a likely error of around 30 miles per second. This measurement, made by timing a flash of light travelling between mirrors in Annapolis, agreed well with less direct measurements based on astronomical observations. Still, this did not really clarify the *nature* of light.

Two hundred years earlier, Newton had suggested that light consists of tiny *particles* generated in a hot object, which spray out at very high speed, bounce off other objects, and are detected by our eyes. Newton's arch-enemy Robert Hooke, on the other hand, thought that light must be a kind of *wave motion*, like sound. To appreciate his point of view, let us briefly review the nature of sound.

The Wavelike Nature of Sound

Actually, *sound* was already quite well understood by the ancient Greeks. The essential point they had realized is that sound is generated by a vibrating material object, such as a bell, a string or a drumhead. Their explanation was that the vibrating drumhead, for example, alternately pushes and pulls on the air directly above it, sending out waves of compression and decompression (known as rarefaction), like the expanding circles of ripples from a disturbance on the surface of a pond.

On reaching the ear, these waves push and pull on the eardrum with the same frequency (that is to say, the same number of pushes per second) as the original source was vibrating at, and nerves transmit from the ear to the brain both the intensity (loudness) and frequency (pitch) of the sound.

There are a couple of special properties of sound waves (actually any waves) worth mentioning at this point. The first is called *interference*. This is most simply demonstrated with water waves. If you put two fingers in a tub of water, just touching the surface a foot or so apart, and vibrate them at the same rate to get two expanding circles of ripples, you will notice that where the ripples overlap there are quite complicated patterns of waves formed.

The essential point is that at those places where the wave-crests from the two sources arrive at the same time, the waves will work together and the water will be very disturbed, but at points where the crest from one source arrives at the same time as the wave trough from the other source, the waves will cancel each other out, and the water will hardly move. You can hear this effect for sound waves by playing a constant note through stereo speakers.

As you move around a room, you will hear quite large variations in the intensity of sound. Of course, reflections from walls complicate the pattern. This large variation in volume is *not* very noticeable when the stereo is playing music, because music is made up of many frequencies, and they change all the time. The different frequencies, or notes, have their quiet spots in the room in different places. The other point that should be mentioned is that high frequency tweeter-like sound is much more *directional* than low frequency woofer-like sound. It really doesn't matter where in the room you put a low-frequency woofer—the sound seems to be all around you anyway.

On the other hand, it is quite difficult to get a speaker to spread the high notes in all directions. If you listen to a cheap speaker, the high notes are loudest if the speaker is pointing right at you. A lot of effort has gone into designing tweeters, which are small speakers especially designed to broadcast high notes over a wide angle of directions.

Is Light a Wave

Bearing in mind the above minireview of the properties of waves, let us now reconsider the question of whether light consists of a stream of particles or is some kind of wave. The strongest argument for a particle picture is that light travels in straight lines. You can hear around a corner, at least to some extent, but you certainly can't see. Furthermore, no wave-like interference effects are very evident for light.

Finally, it was long known, as we have mentioned, that sound waves were compressional waves in air. If light is a wave, just what is waving? It clearly isn't just air, because light reaches us from the sun, and indeed from stars, and we know the air doesn't stretch that far, or the planets would long ago have been slowed down by air resistance.

Despite all these objections, it was established around 1800 that light *is* in fact some kind of wave. The reason this fact had gone undetected for so long was that the wavelength is *really* short, about one fifty-thousandth of an inch. In contrast, the shortest wavelength sound detectable by humans has a wavelength of about half an inch.

The fact that light travels in straight lines is in accord with observations on sound that the higher the frequency (and shorter the wavelength) the greater the tendency to go in straight lines. Similarly, the interference patterns mentioned above for sound waves or ripples on a pond vary over distances of the same sort of size as the wavelengths involved. Patterns like that would not normally be noticeable for light because they would be on such a tiny scale. In fact, it turns out, there *are* ways to see interference effects with light. A familiar example is the many colors often visible in a soap bubble. These come about because looking at a soap bubble you see light reflected from both sides of a very thin film of water—a thickness that turns out to be comparable to the wavelength of light.

The light reflected from the lower layer has to go a little further to reach your eye, so that light wave must wave an extra time or two before getting to your eye compared with the light reflected from the top layer. What you actually *see* is the *sum* of the light reflected from the top layer and that reflected from the bottom layer. Thinking of this now as the sum of two sets of waves, the light will be bright if the crests of the two waves arrive together, dim if the *crests* of waves reflected from the top layer arrive simultaneously with the *troughs* of waves reflected from the bottom layer.

Which of these two possibilities actually occurs for reflection from a particular bit of the soap film depends on just how much further the light reflected from the lower surface has to travel to reach your eye compared with light from the upper surface, and that depends on the angle of reflection and the thickness of the film. Suppose now we shine *white* light on the bubble. White light is made up of all the colors of the rainbow, and these different colors have different wavelengths, so we see colors reflected, because for a particular film, at a particular angle, some colors will be reflected brightly (the crests will arrive together), some dimly, and we will see the ones that win.

If Light *is* a Wave, What is Waving

Having established that light is a wave, though, we still haven't answered one of the major objections raised above. Just what is waving? We discussed sound waves as waves of compression in air. Actually, that is only one case—sound will also travel through liquids, like water, and solids, like a steel bar. It is found experimentally that, other things being equal, sound travels faster through a medium that is harder to compress: the material just springs back faster and the wave moves through more rapidly.

For media of equal springiness, the sound goes faster through the less heavy medium, essentially because the same amount of springiness can push things along faster in a lighter material. So when a sound wave passes, the material—air, water or solid—waves as it goes through.

Taking this as a hint, it was natural to suppose that light must be just waves in some mysterious material, which was called the *aether*, surrounding and permeating everything. This aether must also fill all of space, out to the stars, because we can see them, so the medium must be there to carry the light. (We could never *hear* an explosion on the moon, however loud, because there is no air to carry the sound to us).

Let us think a bit about what properties this aether must have. Since light travels so fast, it must be very light, and very hard to compress. Yet, as mentioned above, it must allow solid bodies to pass through it freely, without aether resistance, or the planets would be slowing down. Thus we can picture it as a kind of ghostly wind blowing through the earth. But how can we prove any of this? Can we detect it?

Detecting the Aether Wind: the Michelson-Morley Experiment

Detecting the aether wind was the next challenge Michelson set himself after his triumph in measuring the speed of light so accurately. Naturally, something that allows solid bodies to pass through it freely is a little hard to get a grip on. But Michelson realized that, just as the speed of sound is relative to the air, so the speed of light must be relative to the aether. This must mean, if you could measure the speed of light accurately enough, you could measure the speed of light travelling upwind, and compare it with the speed of light travelling downwind, and the difference of the two measurements should be twice the windspeed. Unfortunately, it wasn't that easy.

All the recent accurate measurements had used light travelling to a distant mirror and coming back, so if there was an aether wind along the direction between the mirrors, it would have opposite effects on the two parts of the measurement, leaving a very small overall effect.

There was no technically feasible way to do a one-way determination of the speed of light. At this point, Michelson had a very clever idea for detecting the aether wind. As he explained to his children, it was based on the following puzzle:

Suppose we have a river of width w (say, 100 feet), and two swimmers who both swim at the same speed v feet per second (say, 5 feet per second). The river is flowing at a steady rate, say 3 feet per second. The swimmers race in the following way: they both start at the same point on one bank. One swims directly across the river to the closest point on the opposite bank, then turns around and swims back. The other stays on one side of the river, swimming upstream a distance (measured along the bank) exactly equal to the width of the river, then swims back to the start. Who wins?

Let's consider first the swimmer going upstream and back. Going 100 feet upstream, the speed relative to the bank is only 2 feet per second, so that takes 50 seconds. Coming back, the speed is 8 feet per second, so it takes 12.5 seconds, for a total time of 62.5 seconds.

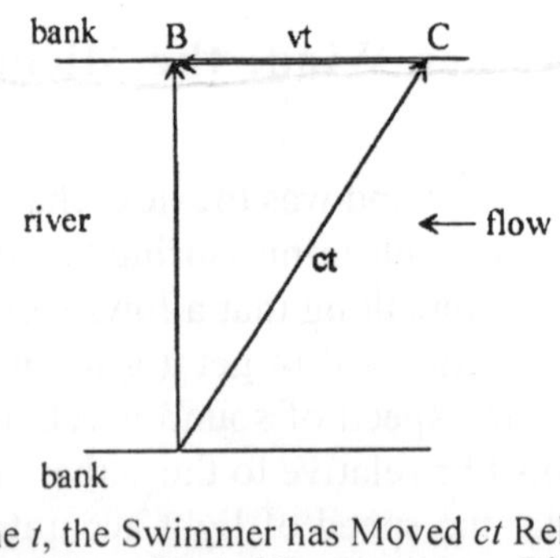

Fig. In time t, the Swimmer has Moved ct Relative to the Water, and been Carried Downstream a Distance vt.

The swimmer going across the flow is trickier. It won't do simply to aim directly for the opposite bank: the flow will carry the swimmer downstream.

To succeed in going directly across, the swimmer must actually aim upstream at the correct angle. Thus, the swimmer is going at 5 feet per second, at an angle, relative to the river, and being carried downstream at a rate of 3 feet per second. If the angle is correctly chosen so that the net movement is directly across, in one second the swimmer must have moved *four feet* across: the distances covered in one second will form a 3, 4, 5 triangle. So, at a crossing rate of 4 feet per second, the swimmer gets across in 25 seconds, and back in the same time, for a total time of 50 seconds. The cross-stream swimmer wins. This turns out to true whatever their swimming speed.

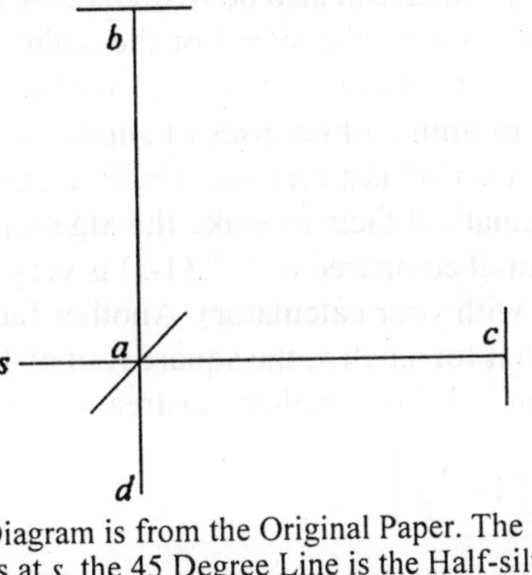

Fig. This Diagram is from the Original Paper. The Source of Light is at *s*, the 45 Degree Line is the Half-silvered Mirror, *b* and *c* are Mirrors and d the Observer.

Michelson's great idea was to construct an exactly similar race for pulses of light, with the aether wind playing the part of the river. The scheme of the experiment is as follows: a pulse of light is directed at an angle of 45 degrees at a half-silvered, half transparent mirror, so that half the pulse goes on through the glass, half is reflected. These two half-pulses are the two swimmers.

They both go on to distant mirrors which reflect them back to the half-silvered mirror. At this point, they are again half reflected and half transmitted, but a telescope is placed behind the half-silvered mirror as shown in the figure so that half of each half-pulse will arrive in this telescope. Now, if there is an aether wind blowing, someone looking through the telescope should see the halves of the two half-pulses to arrive at slightly different times, since one would have gone more upstream and back, one more across stream in general. To maximize the effect, the whole apparatus, including the distant mirrors, was placed on a large turntable so it could be swung around.

Let us think about what kind of time delay we expect to find between the arrival of the two half-pulses of light. Taking the speed of light to be c miles per second relative to the aether, and the aether to be flowing at v miles per second through the laboratory, to go a distance w miles upstream will take $w/(c-v)$ seconds, then to come back will take $w/(c+v)$ seconds. The total roundtrip time upstream and downstream is the sum of these, which works out to be $2wc/(c^2-v^2)$, which can also be written $(2w/c)\text{X}1/(1-v^2/c^2)$. Now, we can safely assume the speed of the aether is much less than the speed of light, otherwise it would have been noticed long ago, for example in timing of eclipses of Jupiter's satellites.

This means v^2/c^2 is a very small number, and we can use some handy mathematical facts to make the algebra a bit easier. First, if x is very small compared to 1, $1/(1-x)$ is very close to $1+x$. (You can check it with your calculator). Another fact we shall need in a minute is that for small x, the square root of $1+x$ is very close to $1+x/2$. Putting all this together, upstream-downstream roundtrip

$$\text{time} \cong \frac{2w}{c} \times \left(1 + \frac{v^2}{c^2}\right)$$

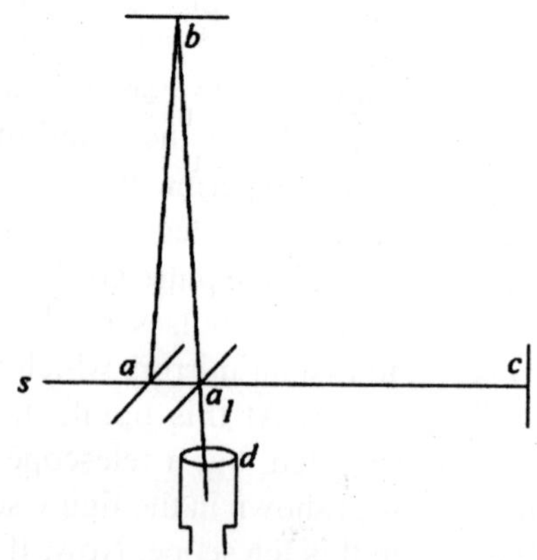

Fig. This is also from the Original Paper, and Shows the Expected Path of Light Relative to the Aether with an Aether wind Blowing.

Now, what about the cross-stream time? The actual cross-stream speed must be figured out as in the example above using a right-angled triangle, with the hypoteneuse equal to the speed c, the shortest side the aether flow speed v, and the other side the cross-stream speed we need to find the time to get across. From Pythagoras' theorem, then, the cross-stream speed is the square root of (c^2-v^2).

Since this will be the same both ways, the roundtrip cross-stream time will be

$$2w/\sqrt{c^2-v^2}.$$

This can be written in the form

$$\frac{2w}{c}\frac{1}{\sqrt{1-v^2/c^2}} \cong \frac{2w}{c}\frac{1}{1-\left(\dfrac{1}{v^2/2c^2}\right)} \cong \frac{2w}{c}\left(1+\frac{v^2}{2c^2}\right)$$

where the two successive approximations, valid for $v/c = x \ll 1$, are $\sqrt{1-x} \cong 1-(x)$ and $1/(1-x) \cong 1 + x$.

Therefore the

$$\text{cross-stream roundtrip time} \cong \frac{2w}{c}\times\left(1+\frac{v^2}{2c^2}\right).$$

Looking at the two roundtrip times at the ends of the two paragraphs above, we see that they differ by an amount $(2w/c)$ X $v^2/2c^2$. Now, $2w/c$ is just the time the light would take if there were no aether wind at all, say, a few millionths of a second. If we take the aether windspeed to be equal to the earth's speed in orbit, for example, v/c is about 1/10, 000, so v^2/c^2 is about 1/100, 000, 000.

This means the time delay between the pulses reflected from the different mirrors reaching the telescope is about one-hundred-millionth of a few millionths of a second. It seems completely hopeless that such a short time delay could be detected. However, this turns out *not* to be the case, and Michelson was the first to figure out how to do it.

The trick is to use the *interference* properties of the lightwaves. Instead of sending pulses of light, Michelson sent in a steady beam of light of a single colour. This can be visualized as a sequence of ingoing waves, with a wavelength one fifty-thousandth of an inch or so. Now this sequence of waves is split into two, and reflected as previously described. One set of waves goes upstream and downstream, the other goes across stream and back. Finally, they come together into the telescope and the eye.

If the one that took longer is half a wavelength behind, its troughs will be on top of the crests of the first wave, they will cancel, and nothing will be seen. If the delay is less than that, there will still be some dimming. However, slight errors in the placement of the mirrors would have the same effect. This is one reason why the apparatus is built to be rotated. On turning it through 90 degrees, the upstream-downstream and the cross-stream waves change places. Now the other one should be behind. Thus, if there is an aether wind, if you watch through the telescope while you rotate the turntable, you should expect to see variations in the brightness of the incoming light.

To magnify the time difference between the two paths, in the actual experiment the light was reflected backwards and forwards several times, like a several lap race.

Michelson calculated that an aether windspeed of only one or two miles a second would have observable effects in this experiment, so if the aether windspeed was comparable to the earth's speed in orbit around the sun, it would be easy to see. In fact, *nothing* was observed. The light intensity did not vary at all. Some time later, the experiment was redesigned so that an aether wind caused by the earth's daily rotation could be detected. Again, nothing was seen.

Finally, Michelson wondered if the aether was somehow getting stuck to the earth, like the air in a below-decks cabin on a ship, so he redid the experiment on top of a high mountain in California. Again, no aether wind was observed. It was difficult to believe that the aether in the immediate vicinity of the earth was stuck to it and moving with it, because light rays from stars would deflect as they went from the moving faraway aether to the local stuck aether.

The only possible conclusion from this series of very difficult experiments was that the whole concept of an all-pervading aether was wrong from the start. Michelson was very reluctant to think along these lines. In fact, new theoretical insight into the nature of light had arisen in the 1860's from the brilliant theoretical work of Maxwell, who had written down a set of equations describing how electric and magnetic fields can give rise to each other. He had discovered that his equations predicted there could be waves made up of electric and magnetic fields, and the speed of these waves, deduced from experiments on how these fields link together, would be 186, 300 miles per second.

This is, of course, the speed of light, so it is natural to assume that light is made up of fast-varying electric and magnetic fields. But this leads to a big problem: Maxwell's equations predict a definite speed for light, and it *is* the speed found by measurements. But what is the speed to be measured relative to? The whole point of bringing in the aether was to give a picture for light resembling the one we understand for sound, compressional waves in a medium. The speed of sound through air is measured relative to air. If the wind is blowing towards you from the source of sound, you will hear the sound sooner. If there isn't an aether, though, this analogy doesn't hold up. So what does light travel at 186, 300 miles per second relative to?

There is another obvious possibility, which is called the emitter theory: the light travels at 186, 300 miles per second relative to the source of the light. The analogy here is between light emitted by a source and bullets emitted by a machine gun. The bullets come out at a definite speed (called the muzzle velocity) relative to the barrel of the gun.

If the gun is mounted on the front of a tank, which is moving forward, and the gun is pointing forward, then relative to the ground the bullets are moving faster than they would if shot from a tank at rest. The simplest way to test the emitter theory of light, then, is to measure the speed of light emitted in the forward direction by a flashlight moving in the forward direction, and see if it exceeds the known speed of light by an amount equal to the speed of the flashlight.

Actually, this kind of direct test of the emitter theory only became experimentally feasible in the nineteen-sixties. It is now possible to produce particles, called neutral pions, which decay each one in a little explosion, emitting a flash of light. It is also possible to have these pions moving forward at 185, 000 miles per second when they self destruct, and to catch the light emitted in the forward direction, and clock its speed.

It is found that, despite the expected boost from being emitted by a very fast source, the light from the little explosions is going forward at the usual speed of 186, 300 miles per second. In the last century, the emitter theory was rejected because it was thought the appearance of certain astronomical phenomena, such as double stars, where two stars rotate around each other, would be affected. Those arguments have since been criticized, but the pion test is unambiguous.

Einstein's Answer

The results of the various experiments discussed above seem to leave us really stuck. Apparently light is not like sound, with a definite speed relative to some underlying medium. However, it is also not like bullets, with a definite speed relative to the source of the light. Yet when we measure its speed we always get the same result. How can all these facts be interpreted in a simple consistent way?

SPECIAL RELATIVITY

Galilean Relativity Again

At this point in the course, we finally enter the twentieth century—Albert Einstein wrote his first paper on relativity in 1905. To put his work in context, let us first review just what is meant by "relativity" in physics. The first example, mentioned in a previous lecture, is what is called "Galilean relativity" and is nothing but Galileo's perception that by observing the motion of objects, alive or dead, in a closed room there is no way to tell if the room is at rest or is in fact in a boat moving at a steady speed in a fixed direction. (You *can* tell if the room is accelerating or turning around).

Everything looks the same in a room in steady motion as it does in a room at rest. After Newton formulated his Laws of Motion, describing how bodies move in response to forces and so on, physicists reformulated Galileo's observation in a slightly more technical, but equivalent, way: they said *the laws of physics are the same in a uniformly moving room as they are in a room at rest.* In other words, the same force produces the same acceleration, and an object experiencing no force moves at a steady speed in a straight line in either case.

Of course, talking in these terms implies that we have clocks and rulers available so that we can actually time the motion of a body over a measured distance, so the physicist envisions the room in question to have calibrations along all the walls, so the position of anything can be measured, and a good clock to time motion. Such a suitably equipped room is called a "*frame of reference*"—the calibrations on the walls are seen as a frame which you can use to specify the precise position of an object at a given time. (This is the same as a set of "coordinates"). Anyway, the bottom line is that no amount of measuring of motions of objects in the "frame of reference" will tell you whether this is a frame at rest or one moving at a steady velocity.

What exactly do we mean by a frame "at rest" anyway? This seems obvious from our perspective as creatures who live on the surface of the earth—we mean, of course, at rest relative to fixed objects on the earth's surface. Actually, the earth's rotation means this isn't quite a fixed frame, and also the earth is moving in orbit at 18 miles per second. From an astronaut's point of view, then, a frame fixed relative to the sun might seem more reasonable.

But why stop there? We believe the laws of physics are good throughout the universe. Let us consider somewhere in space far from the sun, even far from our galaxy. We would see galaxies in all directions, all moving in different ways. Suppose we now set up a frame of reference and check that Newton's laws still work. In particular, we check that the First Law holds—that a body experiencing no force moves at a steady speed in a straight line. *This First law is often referred to as The Principle of Inertia, and a frame in which it holds is called an Inertial Frame.* Then we set up another frame of reference, moving at a steady velocity relative to the first one, and find that Newton's laws are o.k. in this frame too. The point to notice here is that it is not at all obvious which—if either—of these frames is "at rest". We *can*, however, assert that they are both *inertial* frames, after we've checked that in both of them, a body with no forces acting on it moves at a steady speed in a straight line (the speed could be zero). In this situation, Michelson would have said that a frame "at rest" is one at rest relative to the aether. However, his own experiment found motion through the aether to be undetectable, so how would we ever know we were in the right frame? The middle of the nineteenth century there was a substantial advance in the understanding of electric and magnetic fields. (In fact, this advance is in large part responsible for the improvement in living standards since that time).

The new understanding was summarized in a set of equations called Maxwell's equations describing how electric and magnetic fields interact and give rise to each other, just as, two centuries earlier, the new understanding of dynamics was summarized in the set of equations called Newton's laws. The important thing about Maxwell's equations for our present purposes is that they predicted waves made up of electric and magnetic fields that moved at $3X10^8$ meters per second, and it was immediately realized that this was no coincidence—light waves must be nothing but waving electric and magnetic fields. (This is now fully established to be the case).

It is worth emphasizing that Maxwell's work predicted the speed of light from the results of experiments that were not thought at the time they were done to have anything to do with light—experiments on, for example, the strength of electric field produced by waving a magnet.

Maxwell was able to deduce a speed for waves like this using methods analogous to those by which earlier scientists had figured out the speed of sound from a knowledge of the density and the springiness of air.

Generalizing Galilean Relativity to Include Light: Special Relativity

We now come to Einstein's major insight: the Theory of Special Relativity. It is deceptively simple. Einstein first dusted off Galileo's discussion of experiments below decks on a uniformly moving ship, and restated it as:

The Laws of Physics are the same in all Inertial Frames. Einstein then simply brought this up to date, by pointing out that the Laws of Physics must now include Maxwell's equations describing electric and magnetic fields as well as Newton's laws describing motion of masses under gravity and other forces. (*Note for experts and the curious*: we shall find that Maxwell's equations are completely unaltered by special relativity, but, as will become clear later, Newton's Laws do need a bit of readjustment to include special relativistic phenomena.

The First Law is still o.k., the Second Law in the form $F = ma$ is not, because we shall find mass varies; we need to equate force to rate of change of momentum (Newton understood that, of course—that's the way he stated the law!). The Third Law, stated as action equals reaction, no longer holds because if a body moves, its electric field, say, does not readjust instantaneously—a ripple travels outwards at the speed of light. Before the ripple reaches another charged body, the electric forces between the two will be unbalanced.

However, the crucial consequence of the Third Law—the conservation of momentum when two bodies interact, still holds. It turns out that the rippling field itself carries momentum, and everything balances).

Demanding that Maxwell's equations be satisfied in all inertial frames has one major consequence as far as we are concerned. Maxwell's equations give the speed of light to be $3X10^8$ meters per second. Therefore, *demanding that the laws of physics are the same in all inertial frames implies that the speed of any light wave, measured in any inertial frame, must be* $3X10^8$ *meters per second.*

This then is the entire content of the Theory of Special Relativity: the Laws of Physics are the same in any inertial frame, and, in particular, any measurement of the speed of light in any inertial frame will always give $3X10^8$ meters per second.

You *Really* Can't Tell You're Moving

Just as Galileo had asserted that observing gnats, fish and dripping bottles, throwing things and generally jumping around would not help you to find out if you were in a room at rest or moving at a steady velocity, Einstein added that no kind of observation at all, *even measuring the speed of light across your room* to any accuracy you like, would help find out if your room was "really at rest".

This implies, of course, that the concept of being "at rest" is meaningless. If Einstein is right, there is *no* natural rest-frame in the universe. Naturally, there can be no "aether", no thin transparent jelly filling space and vibrating with light waves, because if there were, *it* would provide the natural rest frame, and affect the speed of light as measured in other moving inertial frames.

So we see the Michelson-Morley experiment was doomed from the start. There never was an aether wind. The light was not slowed down by going "upstream"—light *always* travels at the same speed, which we shall now call c,

$$c = 3\text{X}10^8 \text{ meters per second}$$

to save writing it out every time. *This now answers the question of what the speed of light, c, is relative to.* We already found that it is not like sound, relative to some underlying medium. It is also not like bullets, relative to the source of the light (the discredited emitter theory).

Light travels at c relative to the observer, since if the observer sets up an inertial frame (clocks, rulers, etc). to measure the speed of light he will find it to be c. (We always assume our observers are very competent experimentalists!)

Truth and Consequences

The Truth we are referring to here is the seemingly innocuous and plausible sounding statement that all inertial frames are as good as each other—the laws of physics are the same in all of them—and so the *speed of light* is the same in all of them. This Special Theory of Relativity has some surprising consequences, which reveal themselves most dramatically when things are moving at relative speeds comparable to the speed of light. Einstein liked to explain his theory using what he called "thought experiments" involving trains and other kinds of transportation moving at these speeds (technically unachievable so far!), and we shall follow his general approach.

To begin with, let us consider a simple measurement of the speed of light carried out at the same time in two inertial frames moving at half the speed of light relative to each other. The setup is as follows: on a flat piece of ground, we have a flashlight which emits a blip of light, like a strobe. We have two photocells, devices which click and send a message down a wire when light falls on them.

The photocells are placed 10 meters apart in the path of the blip of light, they are somehow wired into a clock so that the time taken by the blip of light to travel from the first photocell to the second, in other words, the time between clicks, can be measured. From this time and the known distance between them, we can easily find the speed of the blip of light.

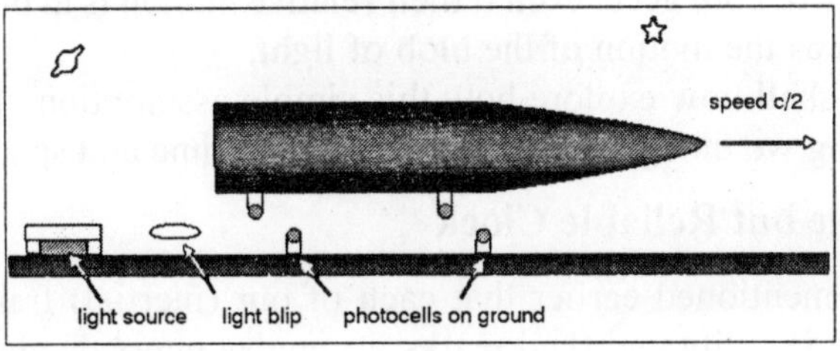

The speed of the same blip of light is measured by two observers, having relative speed c/2. Both measure the time the blip takes from the one photocell to a second one 10 meters further on. Both find the speed to be c.

Meanwhile, there is another observer, passing overhead in a spaceship traveling at half the speed of light. She is also equipped with a couple of photocells, placed 10 meters apart on the bottom of her spaceship as shown, and she is able to measure the speed of the same blip of light, relative to her frame of reference (the spaceship). *The observer on the spaceship will measure the blip of light to be traveling at c relative to the spaceship, the observer on the ground will measure the same blip to be traveling at c relative to the ground.* That is the unavoidable consequence of the Theory of Relativity.

Special Relativity in a Nutshell

Einstein's Theory of Special Relativity, may be summarized as follows:

The Laws of Physics are the same in any Inertial Frame of Reference. (Such frames move at steady velocities with respect to each other).

These Laws include in particular Maxwell's Equations describing electric and magnetic fields, which predict that light always travels at a particular speed c, equal to about $3X10^8$ meters per second, that is, 186, 300 miles per second.

It follows that any measurement of the speed of any flash of light by any observer in any inertial frame will give the same answer c.

We have already noted one counter-intuitive consequence of this, that two different observers moving relative to each other, each measuring the speed of the *same* blob of light relative to himself, will *both* get *c*, even if their relative motion is in the same direction as the motion of the blob of light.

We shall now explore how this simple assumption changes everything we thought we understood about time and space.

A Simple but Reliable Clock

We mentioned earlier that each of our (inertial) frames of reference is calibrated (had marks at regular intervals along the walls) to measure distances, and has a clock to measure time. Let us now get more specific about the clock—we want one that is easy to understand in any frame of reference. Instead of a pendulum swinging back and forth, which wouldn't work away from the earth's surface anyway, we have a blip of light bouncing back and forth between two mirrors facing each other.

We call this device a *light clock*. To really use it as a timing device we need some way to count the bounces, so we position a photocell at the upper mirror, so that it catches the edge of the blip of light. The photocell clicks when the light hits it, and this regular series of clicks drives the clock hand around, just as for an ordinary clock. Of course, driving the photocell will eventually use up the blip of light, so we also need some provision to reinforce the blip occasionally, such as a strobe light set to flash just as it passes and thus add to the intensity of the light. Admittedly, this may not be an easy way to build a clock, but the basic idea is simple.

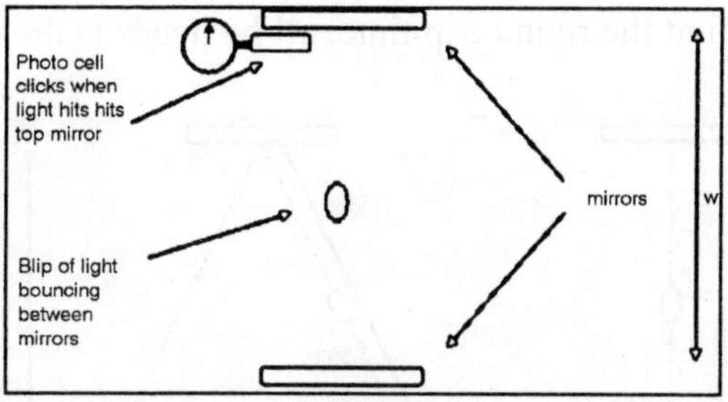

Einstein's light Clock: a blip of light bounces between two parallel mirrors w meters apart; each time it hits the top mirror, the photocell moves the clock on one click. Time between clicks: 2w/c.

It's easy to figure out how frequently our light clock clicks. If the two mirrors are a distance w apart, the round trip distance for the blip from the photocell mirror to the other mirror and back is $2w$. Since we know the blip always travels at c, we find the round trip time to be $2w/c$, so this is the time between clicks.

This isn't a very long time for a reasonable sized clock! The crystal in a quartz watch "clicks " of the order of 10, 000 times a second. That would correspond to mirrors about nine miles apart, so we need our clock to click about 1, 000 times faster than that to get to a reasonable size. Anyway, let us assume that such purely technical problems have been solved.

Looking at Somebody Else's Clock

Let us now consider two observers, Jack and Jill, each equipped with a calibrated inertial frame of reference, and a light clock. To be specific, imagine Jack standing on the ground with his light clock next to a straight railroad line, while Jill and her clock are on a large flatbed railroad wagon which is moving down the track at a constant speed v. Jack now decides to check Jill's light clock against his own. He knows the time for his clock is $2w/c$ between clicks. Imagine it to be a slightly misty day, so with binoculars he can actually see the blip of light bouncing between the mirrors of Jill's clock.

How long does he think that blip takes to make a round trip? The one thing he's sure of is that it must be moving at c = 186, 300 miles per second, relative to him—that's what Einstein tells him. So to find the round trip time, all he needs is the round trip distance.

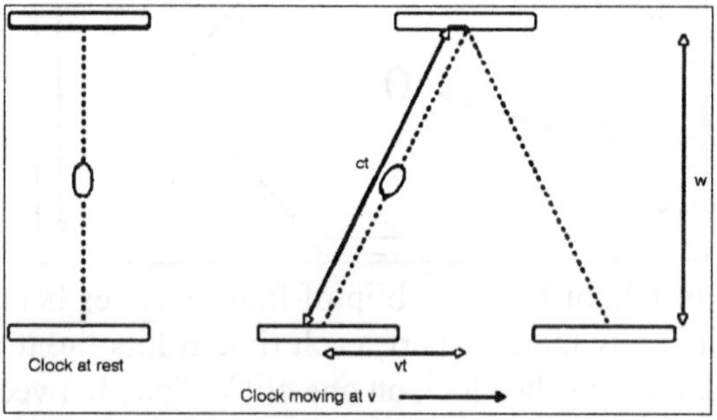

This will *not* be $2w$, because the mirrors are on the flatbed wagon moving down the track, so, relative to Jack on the ground, when the blip gets back to the top mirror, that mirror has moved down the track some since the blip left, so the blip actually follows a zigzag path as seen from the ground.

Two identical light clocks: one at rest, one moving relative to us. The light blips in both travel at the same speed relative to us, the one in the moving clock goes further, so must take longer between clicks, t is the time from one mirror to the other.

Suppose now the blip in Jill's clock on the moving flatbed wagon takes time t to get from the bottom mirror to the top mirror as measured by Jack standing by the track.

Then the length of the "zig" from the bottom mirror to the top mirror is necessarily ct, since that is the distance covered by any blip of light in time t. Meanwhile, the wagon has moved down the track a distance vt, where v is the speed of the wagon.

This should begin to look familiar—it is precisely the same as the problem of the swimmer who swims at speed c relative to the water crossing a river flowing at v! We have again a right-angled triangle with hypotenuse ct, and shorter sides vt and w.

From Pythagoras, then,

$$c^2t^2 = v^2t^2 + w^2$$

so

$$t^2(c^2 - v^2) = w^2$$

or

$$t^2(1 - v^2/c^2) = w^2/c^2$$

and, taking the square root of each side, then doubling to get the round trip time, we conclude that Jack sees the time between clicks for Jill's clock to be:

$$\text{time between clicks for moving clock} = \frac{2w}{c}\frac{1}{\sqrt{1-v^2/c^2}}$$

Of course, this gives the right answer $2w/c$ for a clock at rest, that is, $v = 0$.

This means that Jack sees Jill's light clock to be going slow—a longer time between clicks—compared to his own identical clock. Obviously, the effect is not dramatic at real railroad speeds. The correction factor is $\sqrt{1-v^2/c^2}$, which differs from 1 by about one part in a trillion even for a bullet train! Nevertheless, the effect is real and can be measured.

It is important to realise that the only reason we chose a *light* clock, as opposed to some other kind of clock, is that its motion is very easy to analyse from a different frame. Jill could have a collection of clocks on the wagon, and would synchronize them all. For example, she could hang her wristwatch right next to the face of the light clock, and observe them together to be sure they always showed the same time. Remember, in her frame her light clock clicks every $2w/c$ seconds, as it is designed to do.

Observing this scene from his position beside the track, Jack will see the synchronized light clock and wristwatch next to each other, and, of course, note that the wristwatch is *also* running slow by the factor $\sqrt{1-v^2/c^2}$. In fact, *all* her clocks, including her pulse, are slowed down by this factor according to Jack. Jill is aging more slowly because she's moving!

But this isn't the whole story—we must now turn everything around and look at it from Jill's point of view. *Her inertial frame of reference is just as good as Jack's.* She sees his light clock to be moving at speed v (backwards) so from her point of view *his* light blip takes the longer zigzag path, which means *his clock runs slower than hers.*

That is to say, each of them will see the other to have slower clocks, and be aging more slowly. This phenomenon is called *time dilation*. It has been verified in recent years by flying very accurate clocks around the world on jetliners and finding they register less time, by the predicted amount, than identical clocks left on the ground. Time dilation is also very easy to observe in elementary particle physics.

Fitzgerald Contraction

Consider now the following puzzle: suppose Jill's clock is equipped with a device that stamps a notch on the track once a second. How far apart are the notches? From Jill's point of view, this is pretty easy to answer. She sees the track passing under the wagon at *v* meters per second, so the notches will of course be *v* meters apart.

But Jack sees things differently. He sees Jill's clocks to be running slow, so he will see the notches to be stamped on the track at intervals of $1/\sqrt{1-v^2/c^2}$ seconds (so for a relativistic train going at $v = 0.8c$, the notches are stamped at intervals of 5/3 = 1.67 seconds). Since Jack agrees with Jill that the relative speed of the wagon and the track is *v*, he will assert the notches are not *v* meters apart, but $v/\sqrt{1-v^2/c^2}$ meters apart, a greater distance.

Who is right? It turns out that Jack is right, because the notches are in his frame of reference, so he can wander over to them with a tape measure or whatever, and check the distance. This implies that as a result of her motion, Jill observes the notches to be closer together by a factor $\sqrt{1-v^2/c^2}$ than they would be at rest. This is called the *Fitzgerald contraction*, and applies not just to the notches, but also to the track and to Jack—everything looks somewhat squashed in the direction of motion!

Experimental Evidence for Time Dilation: Dying Muons

The first clear example of time dilation was provided over fifty years ago by an experiment detecting *muons*. These particles are produced at the outer edge of our atmosphere by incoming cosmic rays hitting the first traces of air.

They are unstable particles, with a "half-life" of 1.5 microseconds (1.5 millionths of a second), which means that if at a given time you have 100 of them, 1.5 microseconds later you will have about 50, 1.5 microseconds after that 25, and so on. Anyway, they are constantly being produced many miles up, and there is a constant rain of them towards the surface of the earth, moving at very close to the speed of light. In 1941, a detector placed near the top of Mount

Washington (at 6000 feet above sea level) measured about 570 muons per hour coming in. Now these muons are raining down from above, but dying as they fall, so if we move the detector to a lower altitude we expect it to detect fewer muons because a fraction of those that came down past the 6000 foot level will die before they get to a lower altitude detector. Approximating their speed by that of light, they are raining down at 186, 300 miles per second, which turns out to be, conveniently, about 1, 000 feet per microsecond.

Thus they should reach the 4500 foot level 1.5 microseconds after passing the 6000 foot level, so, if half of them die off in 1.5 microseconds, as claimed above, we should only expect to register about 570/2 = 285 per hour with the same detector at this level. Dropping another 1500 feet, to the 3000 foot level, we expect about 280/2 = 140 per hour, at 1500 feet about 70 per hour, and at ground level about 35 per hour.

To summarize: given the known rate at which these raining-down unstable muons decay, and given that 570 per hour hit a detector near the top of Mount Washington, we only expect about 35 per hour to survive down to sea level. In fact, when the detector was brought down to sea level, it detected about 400 per hour! How did they survive?

The reason they didn't decay is that *in their frame of reference, much less time had passed.* Their actual speed is about 0.994*c*, corresponding to a time dilation factor of about 9, so in the 6 microsecond trip from the top of Mount Washington to sea level, their clocks register only 6/9 = 0.67 microseconds. In this period of time, only about one-quarter of them decay.

What does this look like from the muon's point of view? How do they manage to get so far in so little time? To them, Mount Washington and the earth's surface are approaching at 0.994c, or about 1, 000 feet per microsecond.

But in the 0.67 microseconds it takes them to get to sea level, it would seem that to them sea level could only get 670 feet closer, so how could they travel the whole 6000 feet from the top of Mount

Washington? The answer is the Fitzgerald contraction. To them, Mount Washington is squashed in a vertical direction (the direction of motion) by a factor of $\sqrt{1-v^2/c^2}$,the same as the time dilation factor, which for the muons is about 9. So, to the muons, Mount Washington is only 670 feet high—this is why they can get down it so fast.

SYNCHRONIZING CLOCKS

Suppose we want to synchronize two clocks that are some distance apart.

We could stand beside one of them and look at the other through a telescope, but we'd have to remember in that case that we are seeing the clock *as it was when the light left it*, and correct accordingly.

Another way to be sure the clocks are synchronized, assuming they are both accurate, is to start them together. How can we do that? We could, for example, attach a photocell to each clock, so when a flash of light reaches the clock, it begins running.

If, then, we place a flashbulb at the midpoint of the line joining the two clocks, and flash it, the light flash will take the same time to reach the two clocks, so they will start at the same time, and therefore be synchronized.

Let us now put this whole arrangement - the two clocks and the midpoint flashbulb - on a train, and we suppose the train is moving at some speed v to the right, say half the speed of light or so.

Let's look carefully at the clock-synchronizing operation as seen from the ground. In fact, an observer on the ground would say the clocks are *not* synchronized by this operation! The basic reason is that he would see the flash of light from the middle of the train traveling at *c relative to the ground* in each direction, but he would also observe the back of the train coming at v to meet the flash, whereas the front is moving at v away from the bulb, so the light flash must go further to catch up.

In fact, it is not difficult to figure out how much later the flash reaches the front of the train compared with the back of the train, as viewed from the ground. First recall that as viewed from the ground the train has length $L\sqrt{1-v^2/c^2}$

The train is moving to the right: the central bulb emits a flash of light. Seen from the ground, the part of the flash moving towards the rear travels at c, the rear travels at v to meet it.

Letting t_B be the time it takes the flash to reach the back of the train, it is clear from the figure that from which t_B is given by

In a similar way, the time for the flash of light to reach the front of the train. Therefore the time difference between the starting of the two clocks, as seen from the ground.

Remember, this is the time difference between the starting of the train's back clock and its front clock as measured by an observer on the ground with clocks on the ground. However, to this observer the clocks on the train appear to tick more slowly, by the factor , so that although the ground observer measures the time interval between the starting of the clock at the back of the train and the clock at the front as seconds, he also sees the slow running clock at the back actually reading seconds at the instant he sees the front clock to start.

To summarize: as seen from the ground, the two clocks on the train (which is moving at v in the x-direction) are running slowly, registering only seconds for each second that passes. Equally important, the clocks—which are synchronized by an observer on the train—appear unsynchronized when viewed from the ground, the one at the back of the train reading seconds ahead of the clock at the front of the train, where L is the rest length of the train (the length as measured by an observer on the train).

Note that if $L = 0$, that is, if the clocks are together, both the observers on the train and those on the ground will agree that they are synchronized. We need a *distance* between the clocks, as well as relative motion, to get a disagreement about synchronization.

Chapter 7

The Lorentz Transformations

PROBLEMS WITH THE GALILEAN TRANSFORMATIONS

We have already seen that Newtonian mechanics is invariant under the *Galilean* transformations relating two inertial frames moving with relative speed v in the x-direction,

However, these transformations presuppose that *time* is a well-defined universal concept, that is to say, it's the same time everywhere, and all observers can agree on what time it is. Once we accept the basic postulate of special relativity, however, that the laws of physics, including Maxwell's equations, are the same in all inertial frames of reference, and consequently the speed of light has the same value in all inertial frames, then as we have seen, observers in different frames do *not* agree on whether clocks some distance apart are synchronized.

Furthermore, measurements of moving objects are compressed in the direction of motion by the Lorentz-Fitzgerald contraction effect. Obviously, the above equations are too naove! We must think more carefully about time and distance measurement, and construct new transformation equations consistent with special relativity.

Our aim here, then, is to find a set of equations analogous to those above giving the coordinates of an *event* (x, y, z, t) in frame S, for example, a small bomb explosion, as functions of the coordinates (x , y, z, t) of the same event measured in the parallel frame S' which is moving at speed v along the x-axis of frame S. Observers O at the origin in frame S and O' at the origin in frame S synchronize their clocks at $t = t' =0$, at the instant they pass each other, that is, when the two frames coincide . (Using our previous notation, O is Jack and O' is Jill). To determine the time t' at which the bomb exploded in her frame, O' could determine the distance of the point (x', y', z') from the origin, and hence how long it would take light from the explosion to reach her at the origin. A more direct approach (which is helpful in considering transformations between different frames) is to imagine O' to have a multitude of helpers, with an array of clocks throughout the frame, which have all been synchronized by midpoint flashes as described in the previous lecture. Then the event—the bomb explosion—will be close to a clock, and that local clock determines the time t' of the event, so we do not need to worry about timing a light signal. In frame S', then, O' and her crew have clocks all along the x' -axis and all synchronized:

Now consider how this string of clocks appears as viewed by O from frame S. First, since they are all moving at speed v, they will be registering time more slowly by the usual time dilation factor $\sqrt{1-v^2/c^2}$ than O's own physically identical clocks. Second, they will not be synchronized. From the clocks on a train argument in the last lecture, if the clocks are L apart as measured by O', successive clocks to the right (the direction of motion) will be behind by Lv/c^2 as observed by O.

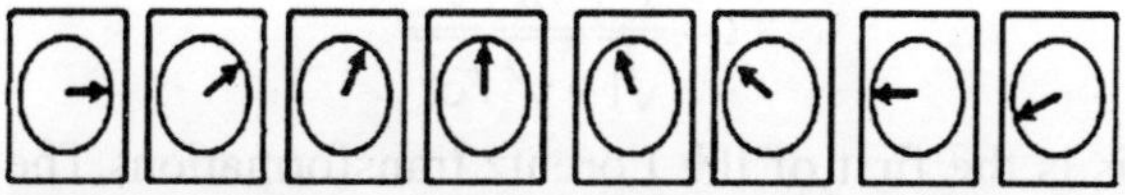

It should be mentioned that this lack of synchronization as viewed from another frame only occurs for clocks *separated in the direction of relative motion.* Consider two clocks some distance apart on the z' axis of S'.

If they are synchronized in S' by both being started by a flash of light from a bulb half way between them, it is clear that as viewed from S the light has to go the same distance to each of the clocks, so they will still be synchronized (although they will start later by the time dilation factor).

Deriving the Lorentz Transformations

Let us now suppose that O' and her crew observe a small bomb to explode in S' at $(x', 0, 0, t')$. The space coordinates and time (x, y, z, t) of this event as observed by O in the frame S. (As above, S' moves relative to S at speed v along the x-axis). In other words, we shall derive the Lorentz transformations—which are just the equations giving the four coordinates of an event in one inertial frame in terms of the coordinates of the same event in another inertial frame. We take y', z' zero because they transform trivially—there is no Lorentz contraction perpendicular to the motion, so $y = y'$ and $z = z'$.

First, we consider at what *time* the bomb explodes as measured by O. O' 's crew found the bomb to explode at time t' as measured by a local clock, that is, one located at the site of the explosion, x'. Now, as observed by O from frame S, O' 's clock at x' is *not* synchronized with O''s clock at the origin S'.

When the bomb explodes and the clock at x' reads t', O will see O''s origin clock to read What does O' s own clock read at this point? Recall that O, O' synchronized their origin clocks at the moment they were together, at Subsequently, O will have observed O''s clock to be running slowly by the time-dilation factor. Therefore, when at the instant of the explosion he sees O''s origin clock to be reading he will find that the true time t in his frame is equal to this appropriately scaled to allow for time dilation, that is,

$$t = \frac{t' + vx'/c^2}{\sqrt{1 - v^2/c^2}}.$$

This is the first of the Lorentz transformations.The second question is: *where* does O observe the explosion to occur?

Since it occurs at time t after O' passed O, O' is vt meters beyond O at the time of the explosion. The explosion takes place x' meters beyond O', as measured by O', but of course O will see that distance x' as contracted to $\sqrt{1 - v^2/c^2}$ since it's in a moving frame.

Therefore O observes the explosion at point x given by

$$x = vt + x'\sqrt{1 - v^2/c^2}.$$

This can be written as an equation for x in terms of x', t' by substituting for t using the first Lorentz transformation above, to give

$$x = \frac{x' + vt'}{\sqrt{1 - v^2/c^2}}.$$

Therefore, we have found the Lorentz transformations expressing the coordinates (x, y, z, t) of an event in frame S in terms of the coordinates (x', y', z', t') of the same event in frame S':

$$x = \frac{x' + vt'}{\sqrt{1 - v^2/c^2}}$$

$$y = y'$$

$$z = z'$$

$$t = \frac{t' + vt'/c^2}{\sqrt{1 - v^2/c^2}}$$

Notice that nothing in the above derivation depends on the x-velocity v of S' relative to S being positive. Therefore, the inverse transformation (from (x, y, z, t) to (x', y', z', t')) has exactly the same form as that given above with v replaced by $-v$.

Spheres of Light

Consider now the following scenario: suppose that as O' passes O (the instant both of them agree is at time $t' = t = 0$) O' flashes a bright light, which she observes to create an expanding spherical shell of light, centered on herself (imagine it' s a slightly foggy day, so she can see how the ripple of light travels outwards). At time t', then, O' (or, to be precise, her local observers out there in the frame) will see a shell of light of radius ct', that is to say, they will see the light to have reached all points (x', y', z') on the surface

$$x'^2 + y'^2 + z'^2 = c^2 t'^2.$$

Question: how do O and his observers stationed throughout the frame S see this light as rippling outwards?

To answer this question, notice that the above equation for where the light is in frame S' at a particular time t' can be written

$$x'^2 + y'^2 + z'^2 = c^2t'^2 = 0.$$

and can be thought of as *a surface in the four dimensional* (x', y', z', t') *space*, the totality of all the "events" of the light reaching any particular point. Now, to find the corresponding surface of events in the four dimensional (x, y, z, t) space, all we have to do is to change from one set of variables to the other using the Lorentz transformations:

On putting these values of (x', y', z', t') into $x'^2 + y'^2 + z'^2 - c^2t'^2 = 0$, we find that the corresponding surface of events in (x, y, z, t) space is:

$$x^2 + y^2 + z^2 - c^2t^2 = 0.$$

This means that at time t, O and his observers in frame S will say the light has reached a spherical surface centered on O.

How can O' and O, as they move further apart, possibly both be right in maintaining that at any given instant the outward moving light pulse has a spherical shape, each saying it is centered on herself or himself?

Imagine the light shell as O' sees it—at the instant t' she sees a sphere of radius r', in particular she sees the light to have reached the spots $+r'$ and $-r'$ on the x' axis. But from O''s point of view the expanding light sphere does *not* reach the point $+r'$ at the same time it reaches $-r'$! (This is just the old story of synchronizing the two clocks at the front and back of the train one more time).

That is why O does not see O''s sphere: *the arrival of the light at the sphere of radius r' around O' at time t' corresponds in S to a continuum of different events happening at different times*.

Lorentz Invariants

We found above that for an event (x', y', z', t') for which the coordinates of the event (x, y, z, t) as measured in the other frame S satisfy The quantity is said to be a *Lorentz invariant*: it doesn't vary on going from one frame to another.

A simple two-dimensional analogy to this invariant is given by considering two sets of axes, Oxy and $Ox'y'$ having the same origin O, but the axis Ox' is at an angle to Ox, so one set of axes is the same as the other set but rotated. The point P with coordinates (x, y) has coordinates (x', y') measured on the $Ox'y'$ axes.

The square of the distance of the point P from the common origin O is $x^2 + y^2$ and is also $x'^2 + y'^2$, so for the transformation from coordinates (x, y) to (x', y'), $x^2 + y^2$ is an invariant. Similarly, if a point P_1 has coordinates (x_1, y_1) and (x_12, y_12) and another point P_2 has coordinates (x_2, y_2) and (x_22, y_22) then clearly the two points are the same distance apart as measured with respect to the two sets of axes, so

$$(x_1 - x_2)^2 + (y_1 - y_2)^2 = (x_1' - x_2')^2 + (y_1' - y_2')^2.$$

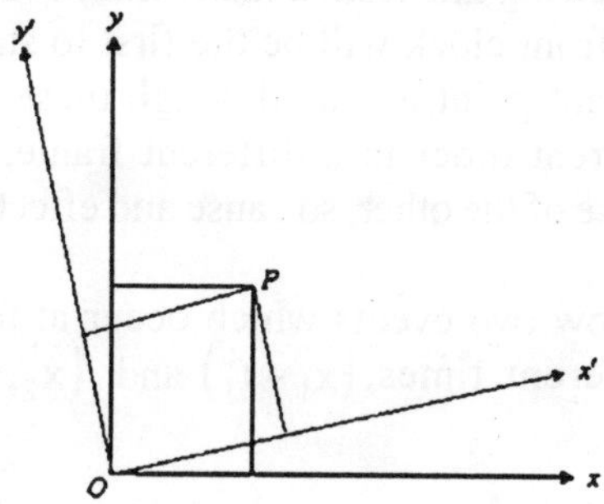

This is really obvious: the distance between two points in an ordinary plane can't depend on the angle at which we choose to set our coordinate axes.

The Lorentz analog of this, dropping the y, z coordinates, can be written

$$c^2(t_1 - t_2)^2 - (x_1 - x_2)^2 = c^2(t_1' - t_2')^2 - (x_1' - x_2')^2 = s^2,$$

say, where s is some sort of measure of the "distance" between the two events (x_1, t_1) and (x_2, t_2).

This s is sometimes called the "space-time interval". *The big difference from the two-dimensional rotation case is that s′can be positive or negative*. For s^2 negative just taking s as its square root the "distance" itself would be imaginary—so we need to think this through carefully. It's clear enough what to do in any particular case—the cases of spacelike and timelike separated events are best dealt with separately, at least to begin with.

Consider first two events simultaneous in frame S', so $t_12 = t_22$. They will not be simultaneous in frame S, but they *will* satisfy

$$(x_1 - x_2)^2 - c^2(t_1 - t_2)^2 > 0.$$

We say the two events are *spacelike separated.* This means that they are sufficiently removed spatially that a light signal could not have time between them to get from one to the other, so one of these events could not be the *cause* of the other. The sequence of two events can be different in different frames if the events are spacelike separated.

Consider again the starting of the two clocks at the front and back of a train as seen from the ground: the back clock starts first. Now imagine viewing this from a faster train overtaking the clock train—now the front clock will be the first to start.

The important point is that although these events appear to occur in a different order in a different frame, neither of them could be the cause of the other, so cause and effect are not switched around.

Consider now two events which occur at the same place in frame *S′*at different times, $(x_1' - t_1')$ and (x_2', t_1') and Then in frame *S*:

$$c^2(t_1 - t_2)^2 - (x_1 - x_2)^2 = c^2(t_1' - t_2')^2 > 0.$$

These events are said to be *timelike* separated. There is no frame in which they are simultaneous. "Cause and effect" events are timelike separated.

The Light Cone

Let us try to visualize the surface in four-dimensional space described by the outgoing shell of light from a single flash,

$$x^2 + y^2 + z^2 - c^2t^2 = 0.$$

It is helpful to think about a simpler situation, the circular ripple spreading on the surface of calm water from a pebble falling in. Taking *c* here to be the speed of the water waves, it is easy to see that at time *t* after the splash the ripple is at

$$x^2 + y^2 - c^2t^2 = 0.$$

Now think about this as a surface in the three-dimensional space (x, y, t). The plane corresponding to time *t* cuts this surface in a circle of radius ct. This means the surface is a cone with its point at the origin. The four-dimensional space flash-of-light surface is not so easy to visualize, but is clearly the higher-dimensional analog: the plane surface corresponding to time t cuts it in a sphere instead of a circle. This surface is called the *lightcone*.

We have stated above that the separation of a point P (x, y, z, t) from the origin is spacelike if $x^2 + y^2 + z^2 - c^2t^2 > 0$ and timelike if $x^2 + y^2 + z^2 - c^2t^2 < 0$. It is said to be *lightlike* if $x^2 + y^2 + z^2 - c^2t^2 = 0$. Points on the light cone described above are lightlike separated from the origin.

To be precise, the points corresponding to an outgoing shell of light from a flash at the origin at $t = 0$ form the *forward light cone*. Since the equation depends only on t^2, there is a solution with t negative, the "backward light cone", just the reflection of the forward light cone in the plane $t = 0$.

Possible causal connections are as follows: an event at the origin (0, 0, 0, 0) could cause an event in the forward light cone: so that is the "future", as seen from the origin. Events in the backward light cone—the "past"—could cause an event at the origin. There can be no causal link between an event at the origin and an event outside the light cones, since the separation is spacelike: outside the light cones is "elsewhere" as viewed from the origin.

TIME DILATION: A WORKED EXAMPLE

Moving Clocks Run Slow plus Moving Clocks Lose Synchronization plus Length Contraction leads to consistency

The object of this exercise is to show explicitly how it is possible for two observers in inertial frames moving relative to each other at a relativistic speed to each see the other's clocks as running slow and as being unsynchronized, and yet if they both look at the same clock at the same time from the same place (which may be far from the clock), they will *agree* on what time it shows!

Suppose that in Jack's frame we have two synchronized clocks C_1 and C_2 set 18×10^8 meters apart (that's about a million miles, or 6 light-seconds). Jill's spaceship, carrying a clock C', is traveling at $0.6c$, that is 1.8×10^8 meters per second, parallel to the line C_1C_2, passing close by each clock. Jill in her relativistic rocket passes jack's first clock at an instant when both their clocks read zero.

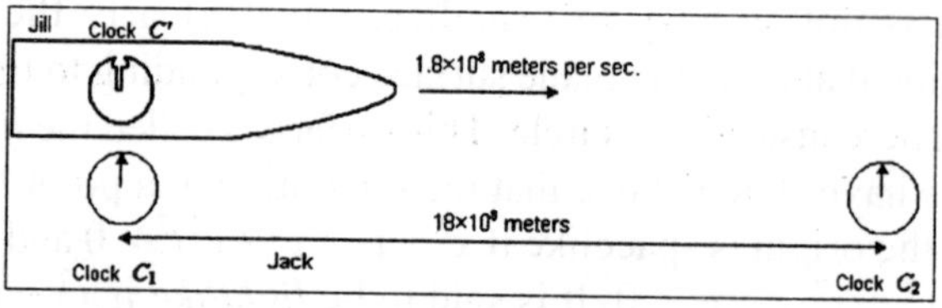

Suppose C' is synchronized with C_1 as they pass, so both read zero.

As measured by Jack the spaceship will take just 10 seconds to reach C_2, since the distance is 6 light seconds, and the ship is traveling at $0.6c$.

What does clock C' (the clock on the ship) read as it passes C_2?

The time dilation factor

$$\sqrt{1-(v^2/c^2)} = 4/5$$

so C', Jill's clock, will read 8 seconds.

Thus if both Jack and Jill are at C_2 as Jill and her clock C' pass C_2, both will agree that the clocks look like:

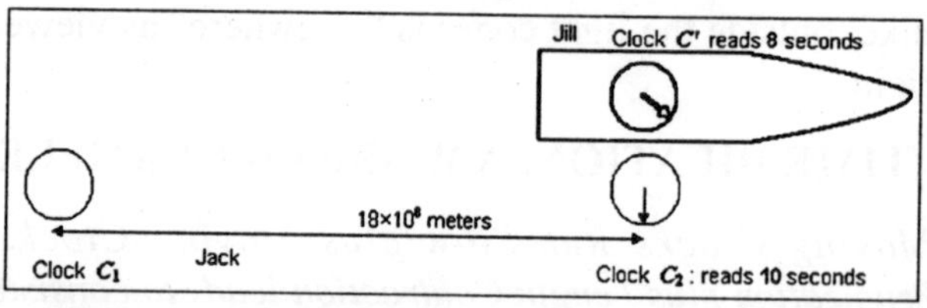

As Jill passes Jack's second clock, both see that his clock reads 10 seconds, hers reads 8 seconds.

How, then, can Jill claim that Jack's clocks C_1, C_2 are the ones that are running slow?

To Jill, C_1, C_2 *are* running slow, but remember they are *not synchronized.* To Jill, C_1 is *behind* C_2 by

Therefore, Jill will conclude that since C_2 reads 10 seconds as she passes it, at that instant C_1 must be registering 6.4 seconds. Jill's own clock reads 8 seconds at that instant, *so she concludes that C_1 is running slow by the appropriate time dilation factor of 4/5.*

This is how the change in synchronization makes it possible for both Jack and Jill to see the other's clocks as running slow. Of course, Jill's assertion that as she passes Jack's second "ground" clock C_2 the first "ground" clock C_1 must be registering 6.4 seconds is not completely trivial to check! After all, that clock is now a million miles away!

Let us imagine, though, that both observers are equipped with Hubble-style telescopes attached to fast acting cameras, so reading a clock a million miles away is no trick.

To settle the argument, the two of them agree that as she passes the second clock, Jack will be stationed at the second clock, and at the instant of her passing they will both take telephoto digital snapshots of the faraway clock C_1, to see what time it reads.

Jack, of course, knows that C_1 is 6 light seconds away, and is synchronized with C_2 which at that instant is reading 10 seconds, so his snapshot must show C_1 to read 4 seconds. That is, looking at C_1 he sees it as it was six seconds ago.

What does Jill's digital snapshot show? It must be identical—two snapshots taken from the same place at the same time must show the same thing! So, Jill *must also* gets a picture of C_1 reading 4 seconds.

How can she reconcile a picture of the clock reading 4 seconds with her assertion that at the instant she took the photograph the clock was registering 6.4 seconds?

The answer is that she can if she knows her relativity!

First point: length contraction. To Jill, the clock C_1 is actually only 4/5 x 18 x 10^8 meters away (she sees the distance C_1C_2 to be Lorentz contracted!).

Second point: The light didn't even have to go that far! In her frame, the clock C_1 is *moving away*, so the light arriving when she's at C_2 must have left C_1 when it was closer—at distance x in the figure below. The figure shows the light in her frame moving from the clock towards her at speed c, while at the same time the clock itself is moving to the left at $0.6c$.

It might be helpful to imagine yourself in her frame of reference, so you are at rest, and to think of clocks C_1 and C_2 as being at the front end and back end respectively of a train that is going past you at speed $0.6c$. Then, at the moment the back of the train passes you, you take a picture (through your telescope, of course) of the clock at the front of the train.

Obviously, the light from the front clock that enters your camera at that instant left the front clock some time ago. During the time that light traveled towards you at speed c, the front of the train itself was going in the opposite direction at speed $0.6c$. But you know the length of the train in your frame is $4/5 \times 18 \times 10^8$ meters, so since at the instant you take the picture the back of the train is passing you, the front of the train must be $4/5 \times 18 \times 10^8$ meters away. Now that distance, $4/5 \times 18 \times 10^8$, is the sum of the distance the light entering your camera traveled plus the distance the train traveled in the same time, that is, $(1 + 0.6)/1$ times the distance the light traveled.

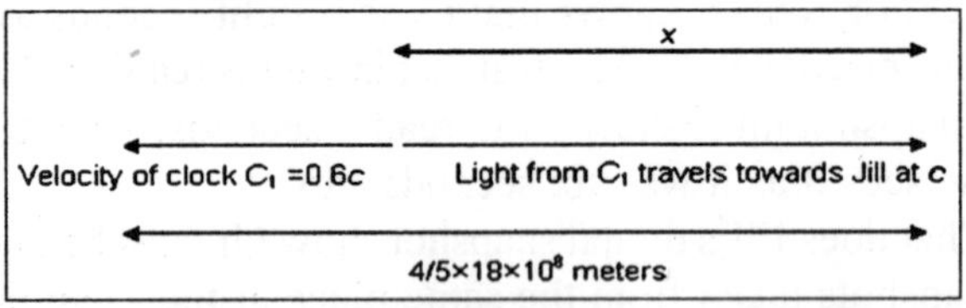

As Jill passes C_2, she phtographs C_1: at that instant, she knows C_1 is $4/5\times18\times10^8$ meters away in her frame, but the light reaching her camera at that moment left C_1 when it was at a distance x, not so far away. As the light traveled towards her at speed c, C_1 was moving away at a speed of 0.6c, so the distance $4/5\times18\times10^8$ meters is the sum of how far the light traveled towards her and how far the clock traveled away from her, both starting at x.

So the image of the first ground clock she sees and records as she passes the second ground clock must have been emitted when the first clock was a distance x from her in her frame, where

$$x(1+3/5) = 4/5\times18\times10^8 \text{ meters, so} x = 9\times10^8 \text{ meters.}$$

Having established that the clock image she is seeing as she takes the photograph left the clock when it was only 9×10^8 meters away, that is, 3 light seconds, she concludes that she is observing the first ground clock as it was three seconds ago.

Third point: time dilation. The story so far: she has a photograph of the first ground clock that shows it to be reading 4 seconds. She knows that the light took three seconds to reach her. So, what can she conclude the clock must actually be registering at the instant the photo was taken? If you are tempted to say 7 seconds, you have forgotten that in her frame, the clock is moving at $0.6c$ and hence *runs slow* by a factor 4/5.

Including the time dilation factor correctly, she concludes that in the 3 seconds that the light from the clock took to reach her, the clock itself will have ticked away 3 X 4/5 seconds, or 2.4 seconds.

Therefore, since the photograph shows the clock to read 4 seconds, and she finds the clock must have run a further 2.4 seconds, she deduces that at the instant she took the photograph the clock must actually have been registering 6.4 seconds, which is what she had claimed all along!

The key point of this lecture is that at first it seems impossible for two observers moving relative to each other to both maintain that the other one's clocks run slow. However, by bringing in the other necessary consequences of the theory of relativity, the Lorentz contraction of lengths, and that clocks synchronized in one frame are out of synchronization in another by a precise amount that follows necessarily from the constancy of the speed of light, the whole picture becomes completely consistent.

THE TRAIN AND THE TWINS

EINSTEIN'S DEFINITION OF COMMON SENSE

Although Einstein's theory of special relativity solves the problem posed by the Michelson-Morley experiment—the nonexistence of an ether—it is at a price. The simple assertion that the speed of a flash of light is always c in any inertial frame leads to consequences that defy common sense.

When this was pointed out somewhat forcefully to Einstein, his response was that common sense is the layer of prejudices put down before the age of eighteen. All our intuition about space, time and motion is based on childhood observation of a world in which no objects move at speeds comparable to that of light. Perhaps if we had been raised in a civilization zipping around the universe in spaceships moving at relativistic speeds, Einstein's assertions about space and time would just seem to be common sense.

The real question, from a scientific point of view, is not whether special relativity defies common sense, but whether it can be shown to lead to a *contradiction*. If that is so, common sense wins. Ever since the theory was published, people have been writing papers claiming it *does* lead to contradictions. The previous lecture, the worked example on time dilation, shows how careful analysis of an apparent contradiction leads to the conclusion that in fact there was no contradiction after all. In this lecture, we shall consider other apparent contradictions and think about how to resolve them. This is the best way to build up an understanding of relativity.

Trapping a Train in a Tunne

One of the first paradoxes to be aired was based on the Fitzgerald contraction. Recall that any object moving relative to an observer will be seen by that observer to be contracted, foreshortened in the direction of motion by the ubiquitous factor $\sqrt{1-v^2/c^2}$ .Einstein lived in Switzerland, a very mountainous country where the railroads between towns often go through tunnels deep in the mountains.

Suppose a train of length L is moving along a straight track at a relativistic speed and enters a tunnel, also of length $L\sqrt{1-v^2/c^2}$. There are bandits inhabiting the mountain above the tunnel. They observe a short train, one of length so they wait until this short train is completely inside the tunnel of length L, then they close doors at the two ends, and the train is trapped fully inside the mountain. Now look at this same scenario from the point of view of someone on the train. He sees a train of length L, approaching a tunnel of length $\sqrt{1-v^2/c^2}$, so the tunnel is not as long as the train from his viewpoint! What does he think happens when the bandits close both the doors?

The Tunnel Doors are Closed Simultaneously

The key to understanding what is happening here is that we said the bandits closed the two doors at the ends of the tunnel *at the same time*. How could they arrange to do that, since the doors are far apart? They could use walkie-talkies, which transmit radio waves, or just flash a light down the tunnel, since it's long and straight. Remember, though, that the train is itself going at a speed close to that of light, so they have to be quite precise about this timing!

The simplest way to imaging them synchronizing the closings of the two doors is to assume they know the train's timetable, and at a prearranged appropriate time, a light is flashed halfway down the tunnel, and the end doors are closed when the flash of light reaches the ends of the tunnel. Assuming the light was positioned correctly in the middle of the tunnel, that should ensure that the two doors close simultaneously.

Or are they

Now consider this door-closing operation from the point of view of someone on the train. Assume he's in an observation car and has incredible eyesight, and there's a little mist, so he actually sees the light flash, and the two flashes traveling down the tunnels towards the two end doors. Of course*, the train is a perfectly good inertial frame*, so he sees these two flashes to be traveling in opposite directions, but *both at c, relative to the train*. Meanwhile, he sees the tunnel itself to be moving rapidly relative to the train. Let us say the train enters the mountain through the "front" door.

The observer will see the door at the other end of the tunnel, the "back" door, to be rushing towards him, and rushing to meet the flash of light. Meanwhile, once he's in the tunnel, the front door is receding rapidly behind him, so the flash of light making its way to that door has to travel further to catch it. So the two flashes of light going down the tunnel in opposite directions do not reach the two doors simultaneously as seen from the train.

The concept of simultaneity, events happening at the same time, is not *invariant as we move from one inertial frame to another*. The man on the train sees the back door close first, and, if it is not quickly reopened, the front of the train will pile into it before the front door is closed behind the train.

Does the Fitzgerald Contraction Work Sideways

The above discussion is based on Einstein's prediction that objects moving at relativistic speed appear shrunken in their direction of motion. How do we know that they're not shrunken in all three directions, i.e. moving objects maybe keep the same shape, but just get smaller? This can be seen *not* to be the case through a symmetry argument, also due to Einstein. Suppose two trains traveling at equal and opposite relativistic speeds, one north, one south, pass on parallel tracks. Suppose two passengers of equal height, one on each train, are standing leaning slightly out of open windows so that their noses should very lightly touch as they pass each other. Now, if *N* (the northbound passenger) sees *S* as shrunken in height, *N*'s nose will brush against *S*'s forehead, say, and *N* will feel *S*'s nose brush his chin.

Afterwards, then, *N* will have a bruised chin (plus nose), *S* a bruised forehead (plus nose). But this is a perfectly symmetric problem, so *S* would say *N* had the bruised forehead, etc. They can both get off their trains at the next stations and get together to check out bruises.

They must certainly be symmetrical! The only *consistent symmetrical solution* is given by asserting that *neither* sees the other to shrink in height (i.e. in the direction perpendicular to their relative motion), so that their noses touch each other. Therefore, the Lorentz contraction *only* operates in the direction of motion, objects get squashed but not shrunken.

How to Give Twins Very Different Birthdays

Perhaps the most famous of the paradoxes of special relativity, which was still being hotly debated in national journals in the fifties, is the twin paradox. The scenario is as follows. One of two twins - the sister—is an astronaut .

She sets off in a relativistic spaceship to alpha-centauri, four light-years away, at a speed of, say, 0.6*c*. When she gets there, she immediately turns around and comes back. As seen by her brother on earth, her clocks ran slowly by the time dilation factor $\sqrt{1-v^2/c^2}$, so although the round trip took 8/0.6 years = 160 months by earth time, she has only aged by 4/5 of that, or 128 months. So as she steps down out of the spaceship, she is 32 months younger than her twin brother.

But wait a minute—how does this look from *her* point of view? She sees the earth to be moving at 0.6*c*, first away from her then towards her. So she must see her brother's clock on earth to be running slow! So doesn't she expect her brother on earth to be the younger one after this trip?

The key to this paradox is that this situation is not as symmetrical as it looks. The two twins have quite different experiences. The one on the spaceship is *not* in an inertial frame during the initial acceleration *and* the turnaround and braking periods. (To get an idea of the speeds involved, to get to 0.6*c* at the acceleration of a falling stone would take over six months). Our analysis of how a clock in one inertial frame looks as viewed from another doesn't work during times when one of the frames isn't inertial - in other words, when one is accelerating.

The Twins Stay in Touch

To try to see just how the difference in ages might develop, let us imagine that the twins stay in touch with each other throughout the trip. Each twin flashes a powerful light once a month, according to their calendars and clocks, so that by counting the flashes, each one can monitor how fast the other one is aging.

The questions we must resolve are: If the brother, on earth, flashes a light once a month, how frequently, as measured by her clock, does the sister see his light to be flashing as she moves away from earth at speed 0.6*c*?

How frequently does she see the flashes as she is *returning* at 0.6*c*?

How frequently does the brother on earth see the flashes from the spaceship?

Once we have answered these questions, it will be a matter of simple bookkeeping to find how much each twin has aged.

Figuring the Observed Time between Flashes

To figure out how frequently each twin observes the other's flashes to be, we will use some results from the previous lecture, on time dilation. In some ways, that was a very small scale version of the present problem. Recall that we had two "ground" clocks only one million miles apart.

As the astronaut, conveniently moving at 0.6*c*, passed the first ground clock, both that clock and her own clock read zero. As she passed the second ground clock, her own clock read 8 seconds and the *first* ground clock, which she photographed at that instant, she observed to read 4 seconds.

That is to say, after 8 seconds had elapsed on her own clock, constant *observation* of the first ground clock would have revealed it to have registered only 4 seconds. (This effect is compounded of time dilation and the fact that as she moves away, the light from the clock is taking longer and longer to reach her). Our twin problem is the same thing, at the same speed, but over a longer time - we conclude that observation of any earth clock from the receding spacecraft will reveal it to be running *at half speed*, so the brother's flashes will be seen at the spacecraft to arrive every two months, by spacecraft time.

Symmetrically, as long as the brother on earth observes his sister's spacecraft to be moving away at 0.6*c*, he will see light from her flashes to be arriving at the earth every two months by earth time. To figure the frequency of her brother's flashes observed as she returns towards earth, we have to go back to our previous example and find how the astronaut traveling at 0.6*c* observes time to be registered by the *second* ground clock, the one she's approaching.

We know that as she passes that clock, it reads 10 seconds and her own clock reads 8 seconds. We must figure out what she would have seen that second ground clock to read had she glanced at it through a telescope as she passed the first ground clock, at which point both her own clock and the first ground clock read zero. But at that instant, the reading she would see on the second ground clock must be the same as would be seen by an observer on the ground, standing by the first ground clock and observing the second ground clock through a telescope. Since the ground observer knows both ground clocks are synchronized, and the first ground clock reads zero, and the second is 6 light seconds distant, it must read -6 seconds if observed at that instant.

Hence the astronaut will observe the second ground clock to progress from -6 seconds to +10 seconds during the period that her own clock goes from 0 to 8 seconds. In other words, she sees the clock she is approaching at 0.6*c* to be running *at double speed*.

Finally, back to the twins. During her journey back to earth, the sister will see the brother's light flashing twice a month.

(Evidently, the time dilation effect does not fully compensate for the fact that each succeeding flash has less far to go to reach her).

We are now ready to do the bookkeeping: first, from the sister's point of view.

What does she see

At 0.6*c*, she sees the distance to alpha-centauri to be contracted by the familiar $\sqrt{1-v^2/c^2} = 0.8$ to a distance of 3.2 light years, which at 0.6*c* will take her a time 5.333 years, or, more conveniently, 64 months. During the outward trip, then, she will see 32 flashes from home, she will see her brother to age by 32 months. Her return trip will also take 64 months, during which time she will see 128 flashes, so over the whole trip she will see $128 + 32 = 160$ flashes. That means she will have seen her brother to age by 160 months or 13 years 4 months.

What does he see

As he watches for flashes through his telescope, the stay-at-home brother will see his sister to be aging at half his own rate of aging as long as he sees her to be moving away from him, then aging at twice his rate as he sees her coming back.

At first glance, this sounds the same as what she sees—but it isn't! The important question to ask is *when* does he see her turn around? To him, her outward journey of 4 light years' distance at a speed of 0.6*c* takes her 4/0.6 years, or 80 months.

BUT he doesn't *see* her turn around until 4 years later, because of the time light takes to get back to earth from alpha-centauri. In other words, he will actually see her aging at half his rate for $80 + 48 = 128$ months, during which time he will see 64 flashes. When he *sees* his sister turn around, she is already more than half way back!

Remember, in his frame the whole trip takes 160 months (8 light years at 0.6*c*) so he will only see her aging at twice his rate during the last 160 - 128 = 32 months, during which period he will see all 64 flashes she sent out on her return trip.

Therefore, by counting the flashes of light she transmitted once a month, he will conclude she has aged 128 months on the trip, which by his clock and calendar took 160 months. So when she steps off the spacecraft 32 months younger than her twin brother, neither of them will be surprised.

The Doppler Effect

The above analysis hinges on the fact that a traveller approaching a flashing light at $0.6c$ will see it flashing at *double* its "natural" rate - the rate observed by someone standing still with the light - and a traveller receding at $0.6c$ from a flashing light will see it to flash at only *half* its natural rate.

This is a particular example of the *Doppler Effect*, first discussed in 1842 by the German physicist Christian Doppler. There is a Doppler Effect for sound waves too. Sound is generated by a vibrating object sending a succession of pressure pulses through the air. These pressure waves are analogous to the flashes of light. If you are approaching a sound source you will encounter the pressure waves more frequently than if you stand still.

This means you will hear a higher frequency sound. If the distance between you and the source of sound is increasing, you will hear a lower frequency. This is why the note of a jet plane or a siren goes lower as it passes you. The details of the Doppler Effect for sound are a little different than those for light, because the speed of sound is not the same for all observers - it's 330 meters per second relative to the air.

It isn't difficult to find the general formula for the Doppler shift, that is, the change in frequency observed when the source of waves (or periodic signals in general) is moving. For example, consider a light flashing once a second as observed by someone in the same frame as the light. Let us imagine the light to be attached to a spaceship, passing us at a relativistic speed v, and imagine we see a flash at the instant it passes us. When will we see the next flash?

First, the spaceship's clock is running slow according to us, so it will take $\sqrt{1-v^2/c^2}$ seconds before it emits the next flash. But it's also moving away from us, so we won't see that next flash until the light has traveled back to us over the distance covered by the spaceship between flashes. Since the spaceship is traveling at v, and the time between flashes as measured in our frame is $1/\sqrt{1-v^2/c^2}$, the distance the spaceship covers between flashes measures in our frame is $v/c\sqrt{1-v^2/c^2}$. Since the light coming back to us from the flash is traveling at *c*, it covers this distance in time $\sqrt{1-v^2/c^2}$.

Thus the total time between our observing the first flash as the spaceship passes close by us and the second flash emitted one second later by the spaceship clock is:

$$\frac{1}{\sqrt{1-v^2/c^2}}+\frac{v}{c\sqrt{1-v^2/c^2}}=\sqrt{\frac{1+v/c}{1-v/c}}.$$

The change in frequency is the inverse of the change in time between flashes. Also, notice that the shift for a spaceship *approaching* at speed *v* is the *inverse* of that for one *receding* at speed *v*. We already established that above for the special case of $v = 0.6c$, where we found the frequency halved for the spaceship receding, doubled for it approaching.

There are some significant differences between the Doppler effect for light and that for sound. The most obvious one is that if something is approaching you at a speed higher than the speed of sound, you won't hear a thing until it hits. Nothing can approach you at greater than the speed of light. Another difference is that the Doppler shift for light depends only on the velocity of the emitter relative to that of the receiver. For sound, the shift depends on the velocities of both of them relative to the air.

Finally, consider a signal from a moving object moving along a straight line, but not towards you. For example, consider a train moving along a straight track, and the nearest point on the track is one mile from where you're standing. What frequency do you hear (or see) for the signal emitted when the train was at the nearest point to you on the track?

For a sound signal, you would hear no Doppler shifting at this point. For a light signal, however, you *would* see a frequency shift downwards equal to the time dilation factor. This is called the *transverse* Doppler shift, and is famous historically because it was first detected in 1938 by two experimenters, Ives and Stillwell, who even then didn't believe in special relativity! They interpreted their result as the slowing down of a clock as it moved through the aether.

An important astronomical application of the Doppler Effect is the *red shift*. The light from very distant galaxies is redder than the light from similar galaxies nearer to us. This is because the further away a galaxy is, the faster it is moving away from us, as the Universe expands.

The light is redder because red light is low frequency light (blue is high) and we see low frequency light for the same reason that the astronaut receding from earth sees flashes less frequently. In fact, the farthest away galaxies we can see are receding faster than the $0.6c$ of our astronaut.

ADDING VELOCITIES: A WALK ON THE TRAIN

THE FORMULA

If I walk from the back to the front of a train at 3 m.p.h., and the train is traveling at 60 m.p.h., then common sense tells me that my speed relative to the ground is 63 m.p.h. This obvious truth, the simple addition of velocities, follows from the Galilean transformations. Unfortunately, it can't be quite right for high speeds!

We know that for a flash of light going from the back of the train to the front, the speed of the light relative to the ground is exactly the same as its speed relative to the train, not 60 m.p.h. different. Hence it is necessary to do a careful analysis of a fairly speedy person moving from the back of the train to the front as viewed from the ground, to see how velocities *really* add.

We consider our standard train of length L moving down the track at steady speed v, and equipped with synchronized clocks at the back and the front.

The walker sets off from the back of the train when that clock reads zero. Assuming a steady walking speed of u meters per second (relative to the train, of course), the walker will see the front clock to read L/u seconds on arrival there.

How does this look from the ground? Let's assume that at the instant the walker began to walk from the clock at the back of the train, the back of the train was passing the ground observer's clock, and both these clocks (one on the train and one on the ground) read zero. The ground observer sees the walker reach the clock at the front of the train at the instant that clock reads L/u (this is in agreement with what is observed *on* the train—two simultaneous events *at the same place* are simultaneous to all observers), but at this same instant, the ground observer says the train's *back* clock, where the walker began, reads $L/u + Lv/c^2$. (This follows from our previously established result that two clocks synchronized in one frame, in which they are L apart, will be out of synchronization in a frame in which they are moving at v along the line joining them by a time Lv/c^2).

Now, how much time elapses as measured by the ground observer's clock during the walk? At the instant the walk began, the ground observer saw the clock at the back of the train (which was right next to him) to read zero. At the instant the walk ended, the ground observer would say that clock read $L/u + Lv/c^2$, from the paragraph above. But the ground observer would see that clock to be running slow, by the usual time dilation factor: so he would measure the time of the walk on his own clock to be:

$$\frac{L/u + Lv/c^2}{\sqrt{1-\left(v^2/c^2\right)}}.$$

How *far* does the walker move as viewed from the ground? In the time t_W, the train travels a distance vt_W, so the walker moves this distance plus the length of the train. Remember that the train is contracted as viewed from the ground! It follows that the distance covered *relative to the ground* during the walk is:

$$\begin{aligned}
d_W &= vt_W + L\sqrt{1-\left(v^2/c^2\right)} \\
&= v\frac{L/u + Lv/c^2}{\sqrt{1-\left(v^2/c^2\right)}} + L\sqrt{1-\left(v^2/c^2\right)} \\
&= \frac{vL/u + Lv/c^2 + L - L\left(v^2/c^2\right)}{\sqrt{1-\left(v^2/c^2\right)}} \\
&= \frac{L(1+v/u)}{\sqrt{1-\left(v^2/c^2\right)}}.
\end{aligned}$$

The walker's *speed* relative to the ground is simply d_W/t_W, easily found from the above expressions:

$$\frac{d_W}{t_W} = \frac{1 + v/u}{1/u + v/c^2} = \frac{u + v}{1 + uv/c^2}.$$

This is the appropriate formula for adding velocities. Note that it gives the correct answer, $u + v$, in the low velocity limit, and also if u or v equals c, the sum of the velocities is c.

Walking *Across* the Train

Imagine now a rather wide train, of width w, and the walker begins the walk across the train, which is now equipped with clocks on both sides, when the clock where he begins reads $t = 0$. For walking speed $u_{y'}$ (relative to the train, and across the train is the y-direction) when he reaches the clock at the other side it will read $w/u_{y'}$. How is this seen from the ground?

The width of the train w will be the same, there is no Lorentz contraction in the y-direction for motion in the x-direction. The beginning and ending clocks will also be synchronized as seen from the ground, since they are separated in the y-direction but not the x-direction. However, they are clocks moving at relativistic speed, so they will exhibit the familiar time dilation factor. That is, when they read $w/u_{y'}$, a clock on the ground will read

$$\frac{w}{u_Y} \frac{1}{\sqrt{1 - v^2/c^2}}.$$

Thus, as observed from the ground, walking directly across the train is slowed down by the time dilation factor, just as is every other activity on the train as seen from the ground.

However, for steady motion on the train in an arbitrary direction, velocity components ($u_{x'}$, $u_{y'}$) the cross-train velocity transforms in a more complicated way, because the train clocks at the beginning and end of the walk are now separated in the x-direction, so if they register an elapsed time of w/u_y a ground observer would add a lack of synchronicity term

$$\frac{Lv}{c^2} = \frac{wu_{x'}}{u_y} \frac{v}{c^2}.$$

Thus the time for the walk as observed from the ground

$$t_w = \left(\frac{w}{u_y} + \frac{wu_{x'}}{u_{y'}}\frac{v}{c^2}\right) / \sqrt{1 - v^2/c^2} = \frac{w}{u_y}.$$

From this we find the general formula for transformation of transverse velocities:

$$u_y = u_y \frac{\sqrt{1 - v^2/c^2}}{1 + u_{x'}v/c^2}.$$

For the special case of walking directly across the train, $u_{x'} = 0$, we recover the earlier result, that transverse velocity is simply slowed by the time dilation effect.

Testing the Addition of Velocities Formula

Actually, the first test of the addition of velocities formula was carried out in the 1850s! Two French physicists, Fizeau and Foucault, measured the speed of light in water, and found it to be c/n, where n is the refractive index of water, about 1.33. (This was the result predicted by the wave theory of light).

They then measured the speed of light (relative to the ground) in *moving* water, by sending light down a long pipe with water flowing through it at speed v. They discovered that the speed relative to the ground was not just $v + c/n$, but had an extra term, $v + c/n - v/n^2$.

Their (incorrect) explanation was that the light was a complicated combination of waves in the water and waves in the aether, and the moving water was only partially dragging the aether along with it, so the light didn't get the full speed v of the water added to its original speed c/n.

The true explanation of the extra term is much simpler: velocities don't simply add. To add the velocity v to the velocity c/n, we must use the addition of velocities formula above, which gives the light velocity relative to the ground to be:

$$(v + c/n)/(1 + v/nc)$$

Now, v is much smaller than c or c/n, so $1/(1 + v/nc)$ can be written as $(1 - v/nc)$, giving:

$$(v + c/n)(1 - v/nc)$$

Multiplying this out gives $v + c/n - v/n^2 - v/n \times v/c$, and the last term is smaller than v by a factor v/c, so is clearly negligible.

Therefore, the 1850 experiment looking for "aether drag" in fact confirms the relativistic addition of velocities formula! Of course, there are many other confirmations.

Relativistic Dynamics

The first coherent statement of what physicists now call *relativity* was Galileo's observation almost four hundred years ago that if you were in a large closed room, you could not tell by observing how things move-living things, thrown things, dripping liquids-whether the room was at rest in a building, say, or below decks in a large ship moving with a steady velocity.

More technically (but really saying the same thing!) we would put it that *the laws of motion are the same in any inertial frame.* That is, these laws really only describe *relative* positions and velocities. In particular, they do not single out a special inertial frame as the one that's "really at rest". This was later all written down more formally, in terms of *Galilean transformations.* Using these simple linear equations, motion analyzed in terms of positions and velocities in one inertial frame could be translated into any other. When, after Galileo, Newton wrote down his Three Laws of Motion, they were of course invariant under the Galilean transformations, and valid in any inertial frame.

About two hundred years ago, it became clear that light was not just a stream of particles (as Newton had thought) but manifested definite *wavelike* properties. This led naturally to the question of what, exactly, was waving, and the consensus was that space was filled with an *aether*, and light waves were ripples in this all-pervading aether analogous to sound waves in air.

Maxwell's discovery that the equations describing electromagnetic phenomena had wavelike solutions, and predicted a speed which coincided with the measured speed of light, suggested that electric and magnetic fields were stresses or strains in the aether, and Maxwell's equations were presumably only precisely correct in the frame in which the aether was at rest. However, very precise experiments which should have been able to detect this aether all failed.

About a hundred years ago, Einstein suggested that maybe *all* the laws of physics were the same in *all* inertial frames, generalizing Galileo's pronouncements concerning motion to include the more recently discovered laws of electricity and magnetism. This would imply there could be no special "really at rest" frame, even for light propagation, and hence no aether. This is a very appealing and very simple concept: the same laws apply in all frames. What could be more reasonable? Though, it turns out to clash with some beliefs about space and time deeply held by everybody encountering this for the first time. The central prediction is that since the speed of light follows from the laws of physics (Maxwell's equations) and some simple electrostatic and magnetostatic experiments, which are clearly frame-independent, *the speed of light is the same in all inertial frames*.

That is to say, the speed of a *particular flash* of light will always be measured to be 3 10^8 meters per second even if measured by different observers moving rapidly relative to each other, where each observer measures the speed of the flash relative to himself. Nevertheless, experiments have show again and again that Einstein's elegant insight is right, and everybody's deeply held beliefs are wrong.

The *kinematical* consequences of Einstein's postulate: how measurements of position, time and velocity in one frame relate to those in another, and how apparent paradoxes can be resolved by careful analysis. So far, though, we have not thought much about dynamics.

We know that Newton's Laws of Motion were invariant under the Galilean transformations between inertial frames. We now know that the Galilean transformations are in fact *incorrect* except in the low speed nonrelativistic limit. Therefore, we had better look carefully at Newton's Laws of Motion in light of our new knowledge.

Newton's Laws Revisited

Newton's First Law, the Principle of Inertia, that an object subject to no external forces will continue to move in a straight line at steady speed, is equally valid in special relativity. Indeed, it is the defining property of an inertial frame that this *is* true, and the content of special relativity is transformations between such frames.

Newton's Second Law, stated in the form force = mass x acceleration, cannot be true as it stands in special relativity. This is evident from the formula we derived for addition of velocities. Think of a rocket having many stages, each sufficient to boost the remainder of the rocket (including the unused stages) to $c/2$ from rest.

We could fire them one after the other in a carefully timed way to generate a continuous large force on the rocket, which would get it to $c/2$ in the first firing. If the acceleration continued, the rocket would very soon be exceeding the speed of light. Yet we know from the addition of velocities formula that in fact the rocket never reaches c. Evidently, Newton's Second Law needs updating.

Newton's Third Law, action = reaction, also has problems. Consider some attractive force between two rapidly moving bodies. As their distance apart varies, so does the force of attraction. We might be tempted to say that the force of A on B is the opposite of the force of B on A, at each instant of time, but that implies *simultaneous* measurements at two bodies some distance from each other, and if it happens to be true in A's inertial frame, it won't be in B's.

Conservation Laws

In nonrelativistic Newtonian physics, the Third Law tells us that two interacting bodies feel equal but opposite forces from the interaction. Therefore from the Second Law, the rate of change of momentum of one of the bodies is equal and opposite to that of the other body, thus the *total* rate of change of momentum of the system caused by the interaction is zero.

Consequently, for any *closed* dynamical system (no outside forces acting) *the total momentum never changes*. This is the *law of conservation of momentum*. It does *not* depend on the details of the forces of interaction between the bodies, only that they be equal and opposite.

The other major dynamical conservation law is the conservation of energy. This was not fully formulated until long after Newton, when it became clear that frictional heat generation, for example, could quantitatively account for the apparent loss of kinetic plus potential energy in actual dynamical systems.

Although these conservation laws were originally formulated within a Newtonian worldview, their very general nature suggested to Einstein that they might have a wider validity. Therefore, as a working hypothesis, he assumed them to be satisfied *in all inertial frames*, and explored the consequences. We follow that approach.

Momentum Conservation on the Pool Table

As a warm-up exercise, let us consider conservation of momentum for a collision of two balls on a pool table. We draw a chalk line down the middle of the pool table, and shoot the balls close to, but on opposite sides of, the chalk line from either end, at the same speed, so they will hit in the middle with a glancing blow, which will turn their velocities through a small angle.

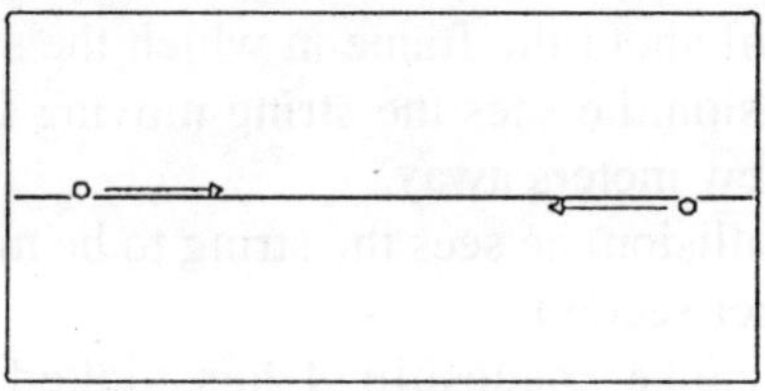

Balls on pool table moving towards glancing collision

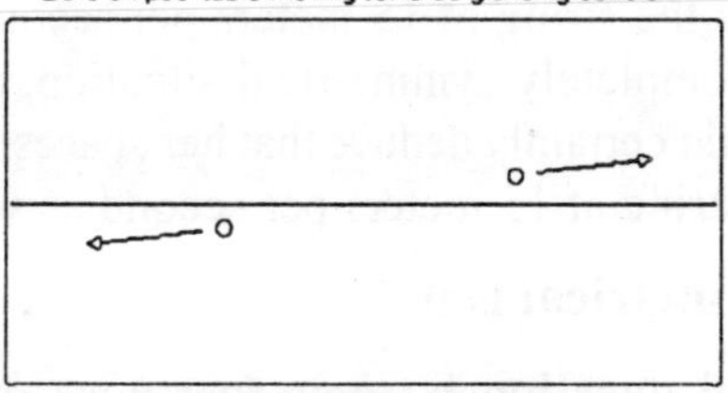

Motion of balls on table after collision

In other words, if initially we say their (equal magnitude, opposite direction) velocities were parallel to the x-direction—the chalk line—then after the collision they will also have equal and opposite small velocities in the y-direction. (The x-direction velocities will have decreased very slightly).

A Symmetrical Spaceship Collision

Now let us repeat the exercise on a grand scale. Suppose somewhere in space, far from any gravitational fields, we set out a string one million miles long. (It could be between our two clocks in the time dilation experiment). This string corresponds to the chalk line on the pool table.

Suppose now we have two identical spaceships approaching each other with equal and opposite velocities parallel to the string from the two ends of the string, aimed so that they suffer a slight glancing collision when they meet in the middle. It is evident from the symmetry of the situation that momentum is conserved in both directions.

In particular, the rate at which one spaceship moves away from the string after the collision - its *y*-velocity - is equal and opposite to the rate at which the other one moves away from the string.

But now consider this collision as observed by someone in one of the spaceships, call it *A*. (Remember, momentum must be conserved in *all* inertial frames—they are all equivalent—there is nothing special about the frame in which the string is at rest). Before the collision, he sees the string moving very fast by the window, say a few meters away.

After the collision, he sees the string to be moving away, at, say, 15 meters per second.

This is because spaceship *A* has picked up a velocity perpendicular to the string of 15 meters per second. Meanwhile, since this is a completely symmetrical situation, an observer on spaceship *B* would certainly deduce that her spaceship was moving away from the string at 15 meters per second as well.

Just how Symmetrical is it

The crucial question is: *how fast does an observer in spaceship A see spaceship B to be moving away from the string?* Let us suppose that relative to spaceship *A*, spaceship *B* is moving away (in the *x*-direction) at $0.6c$.

First, recall that distances perpendicular to the direction of motion are not Lorentz contracted. Therefore, when the observer in spaceship *B* says she has moved 15 meters further away from the string in a one second interval, the observer watching this movement from spaceship *A* will agree on the 15 meters - but disagree on the one second! He will say her clocks run slow, so as measured by his clocks 1.25 seconds will have elapsed as she moves 15 meters in the *y*-direction. It follows that, as a result of time dilation, this collision as viewed from spaceship *A* does *not* cause equal and opposite velocities for the two spaceships in the *y*-direction. Initially, both spaceships were moving parallel to the *x*-axis - there was zero momentum in the *y*-direction.

Consider *y*-direction momentum conservation in the inertial frame in which *A* was initially at rest. An observer in that frame measuring *y*-velocities after the collision will find *A* to be moving at 15 meters per second, *B* to be moving at -0.8 x 15 meters per second in the *y*-direction.

So how can we argue there is zero total momentum in the *y*-direction *after* the collision, when the identical spaceships do *not* have equal and opposite velocities?

Einstein rescues Momentum Conservation

Einstein was so sure that momentum conservation must always hold that he rescued it with a bold hypothesis: the mass of an object must depend on its speed!

In fact, the mass must increase with speed in just such a way as to cancel out the lower *y*-direction velocity resulting from time dilation. That is to say, if an object at rest has a mass m_0, moving at a speed v it must have mass

$$m = \frac{m_0}{\sqrt{1 - v^2 / c^2}}$$

to conserve *y*-direction momentum

Note that this is an undetectably small effect at ordinary speeds, but as an object approaches the speed of light, the mass increases without limit!

Of course, we have taken a very special case here: a particular kind of collision. The reader might well wonder if the same mass correction would work in other types of collision, for example a straight line collision in which a heavy object rear-ends a lighter object.

The algebra is straightforward, if tedious, and it is found that this mass correction factor does indeed ensure momentum conservation for any collision in all inertial frames.

Mass Really *Does* Increase with Speed

Deciding that masses of objects must depend on speed like this seems a heavy price to pay to rescue conservation of momentum! However, it is a prediction that is not difficult to check by experiment. The first confirmation came in 1908, measuring the mass of fast electrons in a vacuum tube. In fact, the electrons in an old-fashioned colour TV tube are about half a percent heavier than electrons at rest, and this must be allowed for in calculating the magnetic fields used to guide them to the screen.

Much more dramatically, in modern particle accelerators very powerful electric fields are used to accelerate electrons, protons and other particles. It is found in practice that these particles become heavier and heavier as the speed of light is approached, and hence need greater and greater forces for further acceleration. Consequently, the speed of light is a natural absolute speed limit. Particles are accelerated to speeds where their mass is thousands of times greater than their mass measured at rest, usually called the "rest mass".

Warning: It should be mentioned that some people don't like the statement that mass increases with speed, they feel that the word "mass" should be restricted to the rest mass of an object, which we've called m_0. This difference of definition has no physical content, however—it's just a matter of taste. We would write momentum as $p = mv$, they would write our m_0 as m, and say the formula for momentum in their notation is p = mn $\sqrt{1-v^2/c^2}$. Either way, a fast electron is that much harder to deflect from a straight line.

Mass and Energy Conservation: Kinetic Energy and Mass for Very Fast Particles

As everyone has heard, in special relativity mass and energy are not separately conserved, in certain situations mass m can be converted to energy $E = mc^2$. This equivalence is closely related to the mass increase with speed. Suppose a constant force F accelerates a particle of rest mass m_0 in a straight line. The work done by the force in accelerating the particle as it travels a distance d is Fd, and this work has given the particle kinetic energy.

As a warm up, recall the elementary derivation of the kinetic energy $\frac{1}{2}mv^2$ of an ordinary *non-relativistic* (i.e. slow moving) object of mass m. Suppose it starts from rest. Then after time t, it has traveled distance $d = \frac{1}{2}at^2$, and $v = at$. From Newton's second law, $F = ma$, the work done by the force $Fd = mad = \frac{1}{2}ma^2t^2 = \frac{1}{2}mv^2$.

This won't work if the mass is varying, because Newton's Second Law isn't always $F = ma$, for variable mass it's

$$F = dp/dt,$$

force = rate of change of momentmum, and if the mass changes the momentum changes, even at constant velocity.

An instructive extreme case is the kinetic energy of a particle traveling close to the speed of light, as particles do in accelerators. In this regime, the change of speed with increasing momentum is negligible! Instead,

$$F = \frac{dp}{dt} = \frac{d(mv)}{dt} \cong \frac{dm}{dt}c$$

where as usual c is the speed of light. This is what happens in a particle accelerator for a charged particle in a constant electric field, with $F = qE$. Since the particle is moving at a speed very close to c, in time dt it will move cdt and the force will do work $Fcdt$. The equation above can be rewritten

$$Fcdt = (dm)c^2.$$

So the energy dE expended by the accelerating force in the time dt yields an increase in mass, and $dE = (dm)c^2$. Provided the speed is close to c, this can of course be integrated to an excellent approximation, to relate a finite particle mass change to the energy expended in accelerating it.

Kinetic Energy and Mass for Slow Particles

Recall that to get momentum to be conserved in all inertial frames, we had to assume an increase of mass with speed by the factor $1/\sqrt{1 - v^2/c^2}$. This necessarily implies that *even a slow-moving object has a tiny mass increase if it is put in motion.*

How does this mass increase relate to the kinetic energy? Consider a mass m_0, moving at speed v, much less than the speed of light. Its kinetic energy $E = \frac{1}{2} m_0 v^2$. Its mass is $m_0 / \sqrt{1 - v^2 / c^2}$, which we can write as $m_0 + dm$, so *dm* is the tiny mass increase we know must occur. It's easy to calculate *dm*.

For n/c << 1, we can make the approximations

$$\sqrt{1 - v^2 / c^2} \cong 1 - \tfrac{1}{2} v^2 / c^2$$

and

$$\frac{1}{1 - \frac{1}{2} v^2 / c} \cong 1 + \tfrac{1}{2} v^2 / c^2.$$

So, for v/c ≪ 1,

$$m(v) \cong m_0 \left(1 + \tfrac{1}{2} v^2 / c^2\right)$$

$$dm(v) \cong \left(\tfrac{1}{2} m_0 v^2\right) / c^2 = KE / c^2.$$

Again, the mass increase *dm* is related to the kinetic energy *KE* by $KE = (dm)c^2$. Having looked at two simple cases, we're ready to derive the general result, valid over the whole range of possible speeds.

Kinetic Energy and Mass for Particles of Arbitrary Speed

We have shown in the two sections above that (in the two limiting cases) when a force does work to increase the kinetic energy of a particle it also causes the mass of the particle to increase by an amount equal to the increase in energy divided by c^2. In fact this result is exactly true over the whole range of speed from zero to arbitrarily close to the speed of light. For a particle of rest mass m_0 accelerating along a straight line (from rest) under a constant force *F*,

$$\begin{aligned} F &= \frac{d}{dt}(mv) \\ &= \frac{dm}{dt} v + m \frac{dv}{dt} \\ &= \frac{m_0}{\left(1 - v^2 / c^2\right)^{3/2}} \frac{v^2}{c^2} \frac{dv}{dt} + \frac{m_0}{\left(1 - v^2 / c^2\right)^{1/2}} \frac{dv}{dt} \\ &= \frac{m_0}{\left(1 - v^2 / c^2\right)^{3/2}} \frac{dv}{dt}. \end{aligned}$$

Therefore, the work done when the particle moves a distance *dx* is

$$Fdx = \frac{m_0}{\left(1 - v^2/c^2\right)^{3/2}} \frac{dv}{dt} dx$$

$$= \frac{m_0}{\left(1 - v^2/c^2\right)^{3/2}} vdv,$$

using $v = dx/dt$.

$$\int Fdx = \int \frac{m_0}{\left(1 - v^2/c^2\right)^{3/2}} vdv = \left(m - m_0\right)c^2.$$

Therefore the total work done from rest—the kinetic energy—is: (The integral is easily done by making the substitution $y = v^2/c^2$). So we see that in the general case the work done on the body, by definition its kinetic energy, is just equal to its mass increase multiplied by c^2.

To understand why this isn't noticed in everyday life, try an example, such as a jet airplane weighing 100 tons moving at 2, 000mph. 100 tons is 100, 000 kilograms, 2, 000mph is about 1, 000 meters per second.

That's a kinetic energy ½ mv^2 of ½.10^{11}joules, but the corresponding mass change of the airplane down by the factor c^2 = 9.10^{16}, giving an actual mass increase of about half a milligram, not too easy to detect!

Notation: m and m_0

As stated earlier, we use m_0 to denote the "rest mass" of an object, and m to denote its relativistic mass, $m = m_0/\sqrt{1 - v^2/c^2}$.

In this notation, we follow French and Feynman. *Krane and Tipler, in contrast, use m for the rest mass*. Using *m* as we do gives neater formulas for momentum and energy, but is not without its dangers. One must remember that *m* is *not* a constant, but a function of speed. Also, *one must remember* that the relativistic kinetic energy is $(m-m_0)c^2$, and *not* equal to ½mv^2, even with the relativistic mass!

Example: take $v^2/c^2 = 0.99$, find the kinetic energy, and compare it with $\frac{1}{2}mv^2$ (using the relativistic mass).

MASS AND ENERGY

Rest Energy

The fact that feeding energy into a body increases its mass suggests that the mass m_0 of a body *at rest*, multiplied by c^2, can be considered as a quantity of energy. The truth of this is best seen in interactions between elementary particles. For example, there is a particle called a *positron* which is exactly like an electron except that it has positive charge. If a positron and an electron collide at low speed (so there is very little kinetic energy) they both disappear in a flash of electromagnetic radiation. This can be detected and its energy measured. It turns out to be $2m_0c^2$ where m_0 is the mass of the electron (and the positron).

Thus particles can "vaporize" into pure energy, that is, electromagnetic radiation. The energy m_0c^2 of a particle at rest is called its "rest energy".

Note, however, that an electron can only be vaporized by meeting with a positron, and there are very few positrons around normally, for obvious reasons-they just don't get far. (Although occasionally it has been suggested that some galaxies may be antimatter!)

Einstein's Box

An amusing "experiment" on the equivalence of mass and energy is the following: consider a closed box with a flashlight at one end and light-absorbing material at the other end. Imagine the box to be far out in space away from gravitational fields or any disturbances. Suppose the light flashes once, the flash travels down the box and is absorbed at the other end.

Now it is known from Maxwell's theory of electromagnetic waves that a flash of light carrying energy E also carries momentum $p = E/c$. Thus, as the flash leaves the bulb and goes down the tube, the box recoils, like a gun, to conserve overall momentum. Suppose the whole apparatus has mass M and recoils at velocity v. Of course, $v \ll c$.

Then from conservation of momentum in the frame in which the box was initially at rest:

$$Mv = E/c,$$

the recoil momentum of the box equals (minus) the momentum of the flash emitted.

After a time $t = L/c$ the light hits the far end of the tube, is absorbed, and the whole thing comes to rest again. (We are assuming that the distance moved by the box is tiny compared to its length).

How far did the box move? It moved at speed v for time t, so it moved distance

$$d = vt = vL/c.$$

From the conservation of momentum equation above, we see that $v = E/Mc$, so the distance d the box moved over is:

$$d = \frac{EL}{mc^2}.$$

Now, the important thing is that there are *no* external forces acting on this system, so *the centre of mass cannot have moved*!

The only way this makes sense is to say that to counterbalance the mass M moving d backwards, the light energy *must have transferred a small mass m*, say, the length L of the tube so that

$$Md = mL$$

and balance is maintained. From our formula for d above, we can figure out the necessary value of m,

$$m = \frac{M}{L}d = \frac{M}{L}\frac{EL}{Mc^2} = \frac{E}{c^2}$$

so

$$E = mc^2.$$

We have therefore established that *transfer of energy implies transfer of the equivalent mass*. Our only assumptions here are that the centre of mass of an isolated system, initially at rest, remains at rest if no external forces act, and that electromagnetic radiation carries momentum E/c, as predicted by Maxwell's equations and experimentally established. But how is this mass transfer physically realized? Is the front end of the tube really heavier after it absorbs the light? The answer is yes, because it's a bit hotter, which means its atoms are vibrating slightly faster—and faster moving objects have higher mass.

Mass and Potential Energy

Suppose now at the far end of the tube we have a hydrogen atom at rest. This atom is essentially a proton having an electron bound to it by electrostatic attraction. It is known that a flash of light with total energy 13.6eV is just enough to tear the electron away, so in the end the proton and electron are at rest far away from each other. The energy of the light was used up dragging the proton and electron apart—that is, it went into potential energy. (It should be mentioned that the electron also loses kinetic energy in this process, 13.6 ev is the net energy required to break up the atom). Now, the light *is* absorbed by this process, so from our argument above the right hand end of the tube *must become heavier*. That is to say, *a proton at rest plus a (distant) electron at rest weigh more than a hydrogen atom* by E/c^2, with E equal to 13.6eV. Thus, Einstein's box forces us to conclude that increased *potential* energy in a system also entails the appropriate increase in mass.

It is interesting to consider the hydrogen atom dissociation in reverse—if a slow moving electron encounters an isolated proton, they may combine to form a hydrogen atom, emitting 13.6eV of electromagnetic radiation energy as they do so. Clearly, then, the hydrogen atom remaining has that much less energy than the initial proton + electron. The actual mass difference for hydrogen atoms is about one part in 10^8.

This is typical of the energy radiated away in a violent chemical reaction—in fact, since most atoms are an order of magnitude or more heavier than hydrogen, a part in 10^9 or 10^{10} is more usual. However, things are very different in nuclear physics, where the forces are stronger so the binding is tighter. A hydrogen nucleus can combine with a lithium nucleus to give two helium nuclei, and the mass shed is 1/500 of the original. This reaction has been observed, and all the masses involved are measurable. The actual energy emitted is 17 MeV. This is the type of reaction that occurs in hydrogen bombs. Notice that the energy released is at least a million times more than the most violent chemical reaction.

As a final example, let's make a ballpark estimate of the change in mass of a million tons of TNT on exploding. The TNT molecule is about a hundred times heavier than the hydrogen atom, and gives off a few eV on burning. So the change in weight is of order $10^{-10} \times 10^6$ tons, about a hundred grams. In a hydrogen bomb, this same mass to energy conversion would take about fifty kilograms of fuel.

Footnote: Einstein's Box is a Fake

Although Einstein's box argument is easy to understand, and gives the correct result, it is based on a physical fiction—the rigid box. If we had a rigid box, or even a rigid stick, all our clock synchronization problems would be over—we could start clocks at the two ends of the stick simultaneously by nudging the stick from one end, and, since it's a rigid stick, the other end would move instantaneously. Actually there are no such materials. All materials are held together by electromagnetic forces, and pushing one end causes a wave of compression to travel down the stick. The electrical forces between atoms adjust at the speed of light, but the overall wave travels far more slowly because each atom in the chain must accelerate for a while before it moves sufficiently to affect the next one measurably.

So the light pulse will reach the other end of the box before it has begun to move! Nevertheless, the wobbling elastic box *does* have a net recoil momentum, which it does lose when the light hits the far end. So the basic point is still valid. French gives a legitimate derivation, replacing the box by its two (disconnected) ends, and finding the centre of mass of this complete system, which of course remains at rest throughout the process.

ENERGY AND MOMENTUM IN LORENTZ TRANSFORMATIONS

HOW DOES THE TOTAL ENERGY OF A PARTICLE DEPEND ON SPEED

We have a formula for the total energy E = K.E. + rest energy,

$$E = mc^2 = \frac{m_0 c^2}{\sqrt{1 - v^2/c^2}}$$

so we can see how total energy varies with speed.

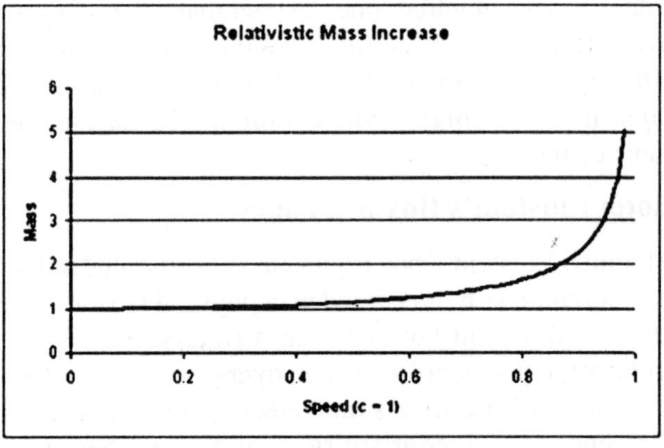

The momentum varies with speed as

$$p = mv = \frac{m_0 v}{\sqrt{1 - v^2/c^2}}.$$

How Does the Total Energy of a Particle Depend on Momentum

It turns out to be useful to have a formula for E in terms of p. Now

$$E^2 = m^2c^4 = \frac{m_0^2 c^4}{1 - v^2/c^2}$$

so

$$m^2c^4(1 - v^2/c^2) = m_0^2c^4$$

$$m^2c^4 - m^2v^2c^2 = m_0^2c^4$$

$$m^2c^4 = E^2 = m_0^2c^4 + m^2c^2v^2$$

hence using $p = mv$ we find

$$E = \sqrt{m_0^2c^2 + c^2p^2}.$$

If p is very small, this gives

$$E \approx m_0 c^2 + \frac{p^2}{2m_0},$$

the usual classical formula.

If p is very large, so $c^2p^2 >> m_0{}^2c^4$, the approximate formula is $E = cp$.

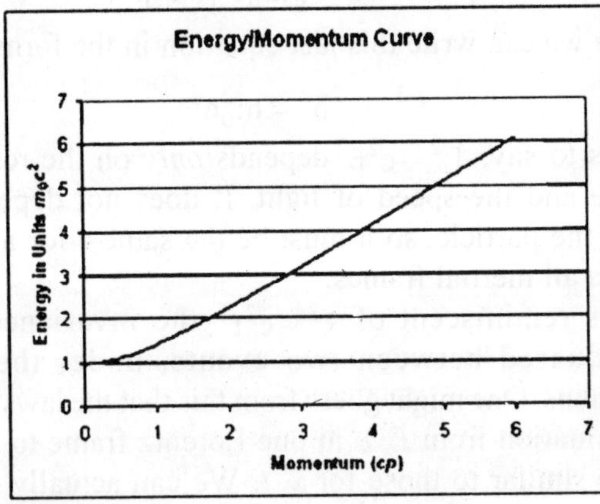

The High Kinetic Energy Limit: Rest Mass Becomes Unimportant

Notice that this high energy limit is just the energy-momentum relationship Maxwell found to be true for light, for all p. This could only be true for *all p* if $m_0{}^2c^4 = 0$, that is, $m_0 = 0$.

Light is in fact composed of "photons"—particles having zero "rest mass". The "rest mass" of a photon is meaningless, since they're never at rest—the energy of a photon

$$E = mc^2 = \frac{m_0 c^2}{\sqrt{1 - v^2/c^2}}$$

is of the form 0/0, since $m_0 = 0$ and $v = c$, so "m" can still be nonzero. That is to say, the mass of a photon is really all K.E. mass. For very fast electrons, such as those produced in high energy accelerators, the additional K.E. mass can be thousands of times the rest mass. For these particles, we can neglect the rest mass and take $E = cp$.

Transforming Energy and Momentum to a New Frame

We have shown

$$\vec{p} = m\vec{v} = \frac{m_0 \vec{v}}{\sqrt{1 - v^2/c^2}}$$

$$E = mc = \sqrt{m_0^2 c^2 + c^2 \vec{p}^2}.$$

Notice we can write this last equation in the form

$$E^2 - c^2\vec{p}^2 = m_0^2 c^4.$$

That is to say, $E^2 - c^2\vec{p}^2$ depends *only* on the rest mass of the particle and the speed of light. It does not depend on the velocity of the particle, so it must be the same—for a particular particle—in all inertial frames.

This is reminiscent of $\vec{x}^2 - c^2t^2$, the invariance of , the interval squared between two events, under the Lorentz transformations. One might guess from this that the laws governing the transformation from *E*, *p* in one Lorentz frame to *E*2 , *p*2 in another are similar to those for *x*, *t*. We can actually derive the laws for *E*, *p* to check this out.

As usual, we consider all velocities to be parallel to the *x*-axis.

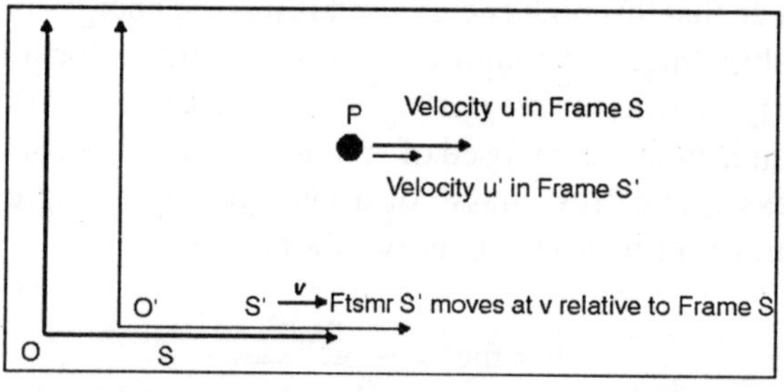

We take the frame *S'* to be moving in the *x*-direction at speed *v* relative to *S*.

Consider a particle of mass m_0 (rest mass) moving at u' in the x' direction in frame S', and hence at *u* along *x* in *S*, where

$$u = \frac{u' + v}{1 + vu'/c^2}.$$

The energy and momentum in *S'* are

$$E' = \frac{m_0c^2}{\sqrt{1-u'^2/c^2}}, p' \frac{m_0u'}{\sqrt{1-u'^2/c^2}}$$

and in *S*:

$$E = \frac{m_0c^2}{\sqrt{1-u^2/c^2}}, p = \frac{m_0u}{\sqrt{1-u^2/c^2}}.$$

Thus

$$E = \frac{m_0c^2}{\sqrt{1-\left(\frac{u'+v}{1+vu'/c^2}\right)/c^2}}$$

giving

$$E = \frac{m_0c^2\left(1+vu'/c^2\right)}{\sqrt{\left(1-v^2/c^2\right)\left(1-u'^2/c^2\right)}}.$$

from which it is easy to show that

$$E = \frac{1}{\sqrt{1-v^2/c^2}}(E'+vp').$$

Similarly, we can show that

$$p = \frac{p'+vE'/c^2}{\sqrt{1-v^2/c^2}}.$$

These are the Lorentz transformations for energy and momentum of a particle—it is easy to check that

$$E^2 - c^2p^2 = E'^2 - c^2p'^2 = m_0^2c^4.$$

Photon Energies in Different Frames

For a zero rest mass particle, such as a photon, $E = cp$, $E^2 - c^2p^2 = 0$ in all frames. Thus

$$E = \frac{E'+vp'}{\sqrt{1-v^2/c^2}} = \frac{E'+vE'/c}{\sqrt{1-v^2/c^2}} = E'\sqrt{\frac{1+v/c}{1-v/c}}.$$

Since $E = cp$, $E2 = cp2$ we also have

$$p = p'\sqrt{\frac{1+v/c}{1-v/c}}.$$

Notice that the ratios of photon *energies* in the two frames coincides with the ratio of photon *frequencies* found in the Doppler shift. When we cover quantum mechanics, the photon energy is proportional to the frequency, so these two must of course transform in identical fashion.

But it's interesting to see it come about this way. Needless to say, relativity gives us no clue on what the constant of proportionality (Planck's constant) is: it must be measured experimentally. But the *same* constant plays a role in all quantum phenomena, not just those concerned with photons.

TRANSFORMING ENERGY INTO MASS: PARTICLE CREATION

PION PRODUCTION

We have mentioned how, using a synchrocyclotron, it is possible to accelerate protons to relativistic speeds. The rest energy of a proton m_pc^2 is 938 MeV, using here the standard high energy physics energy unit: 1 MeV = 10^6 eV. The neutron is a bit heavier—m_nc^2 = 940 MeV. (The electron is 0.51 MeV). Thus to accelerate a proton to relativistic speeds implies giving it a K.E. of order 1, 000 MeV, or 1 GeV.

The standard operating procedure of high energy physicists is to accelerate particles to relativistic speeds, then smash them into other particles to see what happens. For example, fast protons will be aimed at protons at rest (hydrogen atoms, in other words—the electron can be neglected). These proton-proton collisions take place inside some kind of detection apparatus, so the results can be observed.

One widely-used detector is the bubble chamber: a transparent container filled with a superheated liquid. The electric field of a rapidly moving charged particle passing close to a molecule can dislodge an electron, so an energetic particle moving through the liquid leaves a trail of ionized molecules. These give centers about which bubbles can nucleate. The bubbles grow rapidly and provide a visible record of the particle's path. What is actually observed in *p-p* scattering at relativistic energies is that often more particles come out than went in—particles called pions, π^+, π^0, π^- can be created. The π^0 is electrically neutral, the π^+ has exactly the same amount of charge as the proton. It is found experimentally that total electric charge is always conserved in collisions, no matter how many new particles are spawned, and total baryon number (protons + neutrons) is conserved.

Possible scenarios include:

$$p + p \rightarrow p + p + \pi^0$$

and

$$p + p \rightarrow p + n + \pi^+.$$

The neutral pion mass is 135 MeV, the charged pions have mass 140 MeV, where we follow standard high energy practice in calling mc^2 the "mass", since this is the energy equivalent, and hence the energy which, on creation of the particle in a collision, is taken from kinetic energy and stored in mass. However, an incoming proton with 135 MeV of kinetic energy will not be able to create a neutral pion in a collision with a stationary proton. This is because the incoming proton also has momentum, and the collision conserves momentum, so some of the particles after the collision must have momentum and hence kinetic energy.

The simplest way to figure out just how much energy the incoming proton needs to create a neutral pion is to go to the centre of mass frame, where initially two protons are moving towards each other with equal and opposite velocities, there is no total momentum. Obviously, in this frame the least possible K.E. must be just enough to create the π^0 with all the final state particles (*p*, *p*, π^0) at rest. Thus if the relativistic mass of the incoming protons in the centre of mass frame is *m*, the total energy

$$E2mc^2 = 2m_pc^2 + m_zc^2.$$

Putting in the proton and pion masses from above, and using $m = m_p / \sqrt{1-(v^2/c^2)}$, we find the two incoming protons must both be traveling at 0.36*c*.

Recall that this is the speed in the centre of mass frame, and for practical purposes, like designing the accelerator, we need to know the energy necessary in the "lab" frame—that in which one of the protons is initially at rest. The two frames obviously have a relative speed of $0.36c$, so to get the speed of the incoming proton in the lab frame we must add a velocity of $0.36c$ to one of $0.36c$ using the relativistic addition of velocities formula, which gives $0.64c$. This implies the incoming proton has a relativistic mass of 1.3 times its rest mass, and thus a K.E. around 280 MeV.

Thus to create a pion of rest energy 135 MeV, it is necessary to give the incoming proton at least 290 MeV of kinetic energy. This is called the "threshold energy" for pion production. The "inefficiency" arises because momentum is also conserved, so there is still considerable K.E. in the final particles.

Antiproton Production

On raising the energy of the incoming proton further, more particles are produced, including the "antiproton"—a negatively charged heavy particle which will annihilate a proton in a flash of energy. It turns out experimentally that an antiproton can only be produced accompanied by a newly created proton,

$$p + p \rightarrow p + p + p + \overline{p}$$

Notice we could have conserved electric charge with less energy with the reaction

$$p + p \rightarrow p + p + \pi^+ + \overline{p}$$

but this doesn't happen—so charge conserv+ation isn't the only constraint on which particles can be produced. In fact, what we are seeing here is experimental confirmation that the conservation of baryon number, which at the low energies in the context of pion production just meant that the total number of protons plus neutrons stayed fixed, is generalized at high energies to include antiparticles with negative baryon number, -1 for the antiproton. Thus baryon number conservation becomes parallel to electric charge conservation, new particles can always be produced at high enough energies provided the total new charge and the total new baryon number are both zero. We should emphasize again that these are *experimental* results gathered from examining millions of collisions between relativistic particles.

One of the first modern accelerators, built at Berkeley in the fifties, was designed specifically to produce the antiproton, so it was very important to calculate the antiproton production threshold correctly! This can be done by the same method we used above for pion production, but we use a different trick here which is often useful. We have shown that on transforming the energy and momentum of a particle from one frame to another

$$E^2 - c^2p^2 = E'^2 - c^2p'^2$$

Since the Lorentz equations are linear, if we have a system of particles with total energy E and total momentum p in one frame, E', p' in another, it must again be true that

$$E^2 - c^2\vec{p}^2 = E'^2 - c^2\vec{p}'^2.$$

We can use this invariance to get lab frame information from the centre of mass frame. Noting that in the centre of mass (CM) frame the momentum is zero, and in the lab frame the momentum is all in the incoming proton,

$$E_{cm}^2 = ((m_{in} + m_0)c^2)^2 - c^2p_{in}^2$$

where here m_0 is the proton rest mass, m_{in} is the relativistic mass of the incoming proton. At the antiproton production threshold, $E_{cm} = 4m_0c^2$, so

$$16_0^2c^4 = m_{in}^2c^4 + 2m_{in}c^2m_0c + m_0^2c^4 - c^4p_{in}^2$$

and using

$$m_{in}^2c^4 - c^2p_{in}^2 = m_0^2c^4$$

we find

$$2\left(m_{in}c^2\right)\left(m_0c^2\right) + 2\left(m_0c^2\right)^2 = 16\left(m_0c^2\right)^2,$$

so

$$m_{in}c^2 = 7m_0c^2.$$

Therefore to create two extra particles, with total rest energy $2m_0c^2$, it is necessary for the incoming proton to have a kinetic energy of $6m_0c^2$. The Berkeley Gevatron had design energy 6.2 GeV. As we go to higher energies, this "inefficiency" gets worse—consider energies such that the kinetic energy >> rest energy, and assume the incoming particle and the target particle have the same rest mass, m_0, with the incoming particle having relativistic mass m.

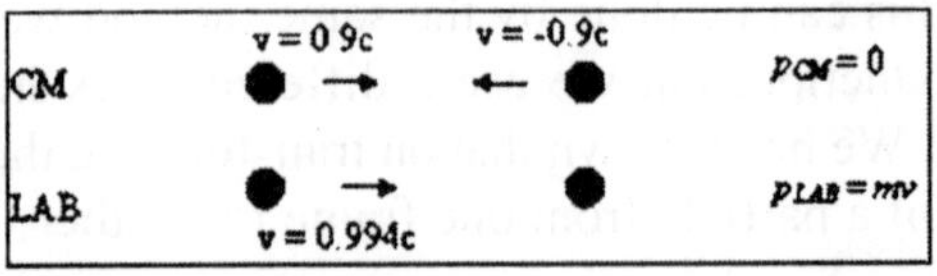

Fig. The Same Collision as Viewed in the CM and LAB Frames

Comparing the centre of mass energy with the lab energy at these high energies,

$$E_{LAB} = (m + m_0)c^2,$$

$$E_{CM}^2 = E_{LAB}^2 - p_{LAB}^2 c^2$$

$$= m^2c^4 + 2mc^2 + m_0^2c^4 - p_{LAB}^2 c^2$$

$$= 2m_0c^2 + (mc^2 + m_0c^2).$$

For $m >> m_0$,

$$E_{CM}^2 \approx 2m_0c^2mc^2 \approx 2m_0c^2.E_{LAB}$$

so

$$E_{CM} \approx \sqrt{2m_0c^2.E_{LAB}},$$

ultimately one must *quadruple* the lab energy to double the centre of mass energy.

Index

A

Accelerated 19, 102, 148, 169, 172, 181, 200, 223, 224, 294
Across the train 270, 271
Adding velocitie 270
Albert Abraham 222
Antiproton production 293
Arbitrary Speed 280
Arrangement 6, 7, 51, 57, 162, 246

B

Boundary 17, 161, 162, 164
Box is a Fake 285

C

Consequences 2, 10, 75, 214, 222, 226, 249, 269, 285
Conservation laws 17, 275
Corresponding States 42

D

Definition of Common 259
Depend on Momentum 286
Depend on speed 278
Deriving the Lorentz 250
Detecting the Aether 227
Detecting the aether 227, 228
Different Birthdays 262
Different frames 172, 248, 249, 254
Doppler effect 112, 123, 125, 128, 277
Dying Muons 244
Dynamics 21, 22, 27, 53, 57, 58, 59, 62, 63, 64, 84, 93, 128, 162, 236, 272, 273

E

Einstein rescues 277
Energy and Momentum 285, 288
Energy and momentum 289, 293
Energy conservation 55
Equivalence Principle 210, 211, 212
Experience to Spacetime 1

F

Fast Flickering 221
Fast Particles 278
Figuring the Observed 263
Fitzgerald contraction 42, 248
Flashes 249, 251, 261, 263, 264, 265, 266, 268, 282
Formula 12, 35, 43, 68, 81, 90,

92, 93, 94, 95, 96, 106, 111, 112, 113, 114, 126, 129, 130, 131, 132, 134, 161, 168, 173, 182, 183, 191, 213, 217, 234, 235, 266, 268, 270, 271, 272
Frames of reference 72, 187, 215, 216, 240

G

Galilean relativity 31, 32, 47, 73, 75, 80, 82, 121, 123, 234
Galilean transformations 33, 35, 53, 62, 75, 84, 217, 248, 268, 272, 273
Generalizing Galilean 236
Global Positioning System 213

I

Inertia and Relativity 19

K

Kinetic energy 279, 280, 281, 284, 291, 293

L

Lanterns 221
Light cone 104, 255
Looking at Somebody 241
Lorentz Invariants 252
Lorentz transformations 48, 76, 98, 138, 146, 254, 292

M

Mass and energy 278, 282
Mass for Slow 279
Mass Really 278
Measuring the speed 220, 227, 238, 240
Messenger 186
Michelson 39, 43, 44, 49, 127, 134, 135, 149, 160, 161, 188, 222, 223, 227, 228, 229, 231, 232, 233, 236, 238, 259
Michelson Measures 222
Momentum 27, 38, 59, 60, 107, 131, 181, 207, 217, 237, 274, 275, 276, 277, 278, 279, 281, 282, 283, 285, 286, 287, 288, 289, 291, 292, 293
More practical 57, 60
Morley 43, 127, 149, 161, 223, 227, 238, 259

N

Nature of light 58, 233
Nature of Sound 224
Newton 15, 20, 22, 26, 29, 36, 69, 186, 188, 223, 234, 237, 272
Null coordinates 101, 104, 106, 107

P

Particle Creation 290
Photon Energies 289, 290
Pion Production 290, 292, 293
Potential energy 275, 284
Problems 85, 241, 248, 285
Propagation 28, 32, 36, 37, 45, 58, 61, 62, 64, 67, 89, 111, 113, 122, 123, 127, 149, 186, 187, 193, 194, 218

R

Reliable Clock 240
Remarks on General 209

Rest energy 290, 292, 293

S

Spaceship Collision 275
Spacetime 1, 11, 27, 39, 45, 51, 61, 74, 84, 85, 86, 89, 91, 93, 94, 98, 101, 104, 105, 106, 107, 108, 112, 136, 139, 155, 158, 170
Special relativity 13, 29, 57, 59, 63, 64, 74, 82, 100, 107, 108, 133, 152, 153, 154, 160, 161, 208, 210, 215, 259, 260, 274, 278
Speed of light 36, 39, 41, 44, 48, 50, 51, 58, 62, 64, 66, 82, 84, 118, 120, 121, 123, 124, 125, 126, 127, 149, 152, 153, 155, 161, 163, 176, 186, 187, 188, 189, 190, 193, 194, 197, 198
Spheres of Light 251
Staircase Wit 74
Starlight and Rain 220
Starry Messenger 186
Stellar Aberration 119
Stellar aberration 120, 121, 125, 126, 127
Symmetrical 27, 28, 81, 88, 89, 90, 95, 98, 106, 156, 262, 263, 264, 275, 276
Synchronizing Clocks 246
Systems of reference 17, 18

T

Thomas precession 200, 201, 206, 207, 208
Time dilation 46, 50, 51, 142, 186, 214, 244, 245, 246, 249, 256, 259, 262, 264, 265, 267, 269, 270, 271, 276
Transformations 10, 11, 12, 26, 33, 39, 41, 48, 62, 70, 71, 74, 75, 76, 81, 82, 87, 91, 98, 136, 138, 142, 146, 152, 248
Transforming Energy 288, 290
Trapping a Train 260
Truth 13, 49, 56, 238, 282

U

Unimportant 287

V

Velocities formula 271, 274

W

Walking 220, 268, 270, 271
Wave 5, 28, 31, 32, 33, 34, 35, 39, 40, 41, 44, 45, 58, 59, 60, 61, 72, 89, 109, 111, 113, 114
Wavelike 35, 59, 60, 224, 272
Work Sideways 261
Worked Example 255, 260